Life Cycle and Development of Diptera

Edited by Muhammad Sarwar

Published in London, United Kingdom

IntechOpen

Supporting open minds since 2005

Life Cycle and Development of Diptera
http://dx.doi.org/10.5772/intechopen.78142
Edited by Muhammad Sarwar

Contributors
Benjamin Kongyeli Badii, Eduardo Jose De Arruda, Márcia Jorge, António Souza, Silvia Martelli, Chaiane Rech, Alexeia Barufatti Grisolia, Bruno Do Amaral Crispim, Ricardo Augusto Dos Passos, Hélina Dos Santos Nascimento, Sahar Abd, Merlin Isaac Kamala, Sonia Kleindorfer, Lauren Common, Rachael Dudaniec, Diane Colombelli-Négrel, Muhammad Sarwar, Valeria Cristina Pinheiro, Aylane Tamara Dos Santos Andrade, Maria Juliana Trindade Bezerra, Joao Zequi, Ediane Amaral, André Silva, William Silva, Karina Silva, Rosemary Roque, Wanderli Tadei, Adrienne Brundage

First published in London, United Kingdom, 2020 by IntechOpen
IntechOpen is the global imprint of INTECHOPEN LIMITED, registered in England and Wales, registration number: 11086078, 5 Princes Gate Court, London, SW7 2QJ, United Kingdom
Printed in Croatia

British Library Cataloguing-in-Publication Data
A catalogue record for this book is available from the British Library

Additional hard and PDF copies can be obtained from orders@intechopen.com

Life Cycle and Development of Diptera
Edited by Muhammad Sarwar
p. cm.
Print ISBN 978-1-83880-225-7
Online ISBN 978-1-83880-226-4
eBook (PDF) ISBN 978-1-83968-738-9

For EU product safety concerns:
IN TECH d.o.o., Prolaz Marije Krucifikse Kozulić 3, 51000 Rijeka, Croatia, info@intechopen.com or visit our website at intechopen.com.

Meet the editor

Dr. Muhammad Sarwar, Principal Scientist, is in his thirtieth year of service with the Department of Agriculture, Government of Punjab. He is also currently working for the Pakistan Atomic Energy Commission. He completed his post doctorate in 2008, funded by the Higher Education Commission of Pakistan from the Institute of Plant Protection in the Chinese Academy of Agricultural Sciences, Beijing, China. He has several hundred published papers to his credit and is recipient of the Shield award, letters of appreciation, and certificates of performance from faculty members of the Chinese Academy of Agricultural Sciences, Beijing, China. In 2010, the Zoological Society of Pakistan presented him with the Prof. Dr. Mirza Azhar Beg Gold Medal. In 2011, the Pakistan Council for Science and Technology awarded him a Research Productivity Award. His research activities focus on integrated pest management for rice, cotton, chickpea, and *Brassica* crops; predatory mites, ladybird beetles, *Chrysoperla*, *Trichogramma*, and parasitoids of fruit flies culturing as bio-control agents; integrated management of fruit flies and mosquitos; and other arthropod pest control methodologies. He has also researched vertebrate pest control, especially controls of rodents in field crops and storage. He was the first to explore thirty-six new species of stored grain mites belonging to eight genera, including *Forcellinia*, *Lackerbaueria*, *Acotyledon*, *Caloglyphus*, and *Troupeauia* in the Acaridae family; and *Capronomoia*, *Histiostoma*, and *Glyphanoetus* in the Histiostomatidae family. He also planned and designed research trials on the integrated management of cotton leaf curl virus (CLCV), pest scouting, pest monitoring, and forecasting. He conducted training of progressive farmers and field staff, and provided advisory services to the farmers regarding plant protection practices. He also trained pesticide dealers on the proper handling, distribution, and storing of pesticides. Under a coordinated research program, Dr. Sarwar collaborated with other institutes to trace resistance sources for cotton, rice, gram, rapeseed, mustard plants, and stored cereals and pulses. He has supervised post-graduate research and is an external examiner for post-graduate studies. He has also organized various workshops, served as a reviewer for scientific journals, and is a member of various working committees. He is responsible for opening up a new avenue on rearing of predatory mites as bio-control agents of insects and mites pests in greenhouse and field crops. Dr. Sarwar is an approved supervisor with the Higher Education Commission (HEC) of Pakistan. He completed a course in Basic Management organized by the Pakistan Institute of Engineering and Applied Sciences (PIEAS), Islamabad, in 2011. He has also completed trainings in Beijing, Bangkok, Havana, and Vienna. His is a lifetime member of the Zoological Society of Pakistan.

Contents

Preface

Diptera, or true flies, are of considerable economic importance. Pestiferous groups can have significant effects on agriculture, animal and human health, and forestry. Other groups can be a general nuisance when present in large numbers or because of allergic reactions due to detached body setae. Despite these negative effects, flies have a valuable role as scavengers, parasitoids and predators of other insects, pollinators, food for predators, bio-indicators of water quality, and tools for scientific research. The basic features of Diptera are the presence of an anterior pair of wings and a posterior pair of club-like balancing organs called halteres.

This book presents a beautiful and useful account of research on Diptera. It is an indispensable resource for entomologists, biologists, and naturalists.

Chapter 1 explores aspects of the biology, ecology, physiology, behavior, taxonomy, and morphology of Diptera. Life histories, habits and habitats, diversity, classification and phylogeny, and detailed keys and diagrams for the correct identification of the more common species of flies are particularly useful sections. All of the world's fly families are included, ranging from common deer flies and fruit flies to deadly tsetse flies and mosquitoes. The illustrated keys linked to the coverage of the world's fly families and subfamilies enable the reader to identify most flies quickly and accurately. This chapter also focuses on strategies for involving the wider community in the control of these pests and development of control programs.

Chapter 2 reviews systematics and taxonomy of the parasitic larvae of *Philornis downsi* Dodge and Aitken (Diptera: Muscidae). It discusses shifts in feeding habits across the species, comparing basal to more recently evolved groups, as well as examines the differences in the ontogeny of wild and captive *P. downsi* larvae, describes what is known about adult *P. downsi* behavior, and discusses changes in *P. downsi* behavior first discovered in Darwin's finch nest.

Chapter 3 on life cycle and cytogenetic study of the mosquito *Culex quinquefasciatus* Say is especially good for evolutionary and genetic studies involving chromosomal polymorphism.

Chapter 4 highlights important information about the yellow fever mosquito *Aedes aegypti* (Diptera: Culicidae) and the identification and quantification of the main breeding sites for this vector. In addition, this chapter compares the characteristics and similarities of the *Aedes* species.

Chapter 5 characterizes the proliferation sites of *Ae. aegypti* in artificial breeding sites in order to examine new strategies for controlling these mosquitoes.

Chapter 6 is on post-embryonic development of *Aedes* (*Stegomyia*) *aegypti* at different temperatures and Co_2 concentrations, and their influences on hatching and development of stabilized population.

Chapter 7 provides evidence and data on the bio-ecology of the blossom midge of jasmine, *Contarinia maculipennis* Felt (Cecidomyiidae) in different jasminum cultivars.

Chapter 8 reviews in detail the current status of Diptera phylogenetics as well as key aspects of the morphology of the different life stages of true flies, which is useful for taxonomic purposes and for understanding the group's biology.

Finally, the last chapter examines information important to forensic entomology, particularly how common flies colonize a dead body. This chapter should be of particular interest to clinicians and forensic entomologists to translate the developmental cycle of flies into usable data for assessing crime scenes.

I am greatly indebted to the all authors who have contributed meaningfully to the success of this work in every regard. I am thankful to each of them for manifold help received at all stages of the work and for materially aiding in the preparation of the book.

My grateful thanks are due to my children Haroon, Farhan, Sidra, Zain-ul-Abideen, and Fatima, as well as my spouse Najma. I also wish to thank my late parents and my brothers and sisters, who are a source of pleasure, inspiration, and encouragement.

All admirations and gratefulness are to almighty Lord of the universe 'Allah' (the one and only God), who is creator, judge, rewarder, unique and inherently one. Allah is the pivot of faith, never sleeps or tires, and perceives and reacts to feelings of every creature and everything through the omnipresence of his divine knowledge. Respect is due to 'Muhammad' who is the last and final greatest Messenger or Prophet in a chain of Messengers and Prophets sent by Allah. Muhammad's sayings, manners, actions, decisions and practices that Muhammad approved, allowed or condoned during lifetime serve as a model for humanity to shape the life in that light to be succcssful in this world and in the life hereafter.

Muhammad Sarwar
National Institute for Biotechnology and Genetic Engineering,
Pakistan

Chapter 1

Typical Flies: Natural History, Lifestyle and Diversity of Diptera

Muhammad Sarwar

Abstract

The Order Diptera, comprising of two-winged or true flies, is one of the most commonly recognized and widespread insects all over the world. During their long evolutionary history, virtually every terrestrial and aquatic niche has been occupied by Diptera, thus making these one of the most successful groups of organisms on earth. The main purpose of this chapter is to provide modern, well-illustrated and easily interpretable information for economic importance, life histories, habits and habitats, lifestyles, diversity, identifying and studying, pharmaceutical and industrial applications, ecological and human services, pests and vectors of diseases, predators and herbivores, pollination and biological control agents, association with carcasses, forensic science, phylogeny and classification of Diptera. Without doubt, this fragment of book provides the basics for understanding diversity of a major order of insects and is the first such synopsis of its kind for scientists and public alike.

Keywords: flies, vectors, pests, ecosystem engineer, forensics

1. Introduction

The so-called true flies are one of the utmost important groups of insects in the order Diptera. The name Diptera, is derived from the Greek words '*di*' meaning two and '*ptera*' hereby meaning wings, which refers to the fact that true flies have only a single pair of wings (two wings). This is distinguishing character because other insects have either two pairs of wings or four wings. Diptera's ancestors also have four wings, but in dipteran insects, second pair of wings is evolved into halteres, which are modified balancer organs that give to insect an amazing amount of fine control while flying.

Many winged insects, such as the butterfly and whitefly, contain the word 'fly' in their names, but are not dipterans, and the name is strictly applicable only to members of Diptera. There is an accepted custom for writing the common names of insects, which have included the word 'fly'. When any name is for a group of insects other than Diptera, it is written as single word (mayfly, dragonfly, and stonefly). But, if an insect belongs to Diptera order and word 'fly' is included, the name is written as double words (horse fly, black fly, crane fly). Diptera is one of the largest insect orders and quite diverse with its numbers more than 125,000 species worldwide. Our world's score of more than 152,000 described species within more than 130 known families is based primarily on figures extracted from the 'BioSystematic Database of World Diptera' [1].

The close association of dipteran flies with humans has led them to be recognized as unpleasant and disturbing creatures, and certainly some flies are responsible for millions of illnesses and deaths among human populations. Yet flies are also among the key components in most ecosystems and known advantageous in many ways. Voluminous flies are of great economic importance because some bloodsuckers are serious pests of humans and other animals. These insects are key vectors of some diseases, although others are pests of cultivated plants. Flies are advantageous as well by operating as predators or parasites of certain insects, scavengers as well as pollinators of plants and killers of weeds harmful to persons. Often called maggots or grubs, dipterous larvae, are found in many habitats (in water, plant tissue and soil, animal matter and decaying plants, below stones or bark, pools of crude petroleum), whereas adults forage on plant or animal juices or other insects. Diptera falls into three big sets, Nematocera (flies with multi-segmented antennae such as crane flies, midges, gnats, mosquitoes), Brachycera (flies with stylate antennae, for instance, horse flies, robber flies, bee flies) and Cyclorrhapha (flies with aristate antennae, such like, flies that breed in vegetable or animal material, both living and dead) [2, 3].

2. General features

Dipteran insects are plentiful all over the sphere, in the tropics and subarctic, at oceanic level and on elevated peaks. These inhabit seashores to low-tide level, however, a small number move into deeper water and merely one or two midges are actually oceanic (*Pontomyia natans* Edwards in the Pacific). In contrast, wandering flies have been observed at much distant to marine. In general, flies vary in dimensions from robber flies more than 7 cm lengthy to midges of little more than 1 mm long. As a whole, the more-primitive flies (midges, mosquitoes, fungus gnats) are fragile insects and with delicate wings. The more-progressive flies (house flies, blow flies) are commonly bristly, thick and tough, and forceful fliers than gnats and midges.

Even though these have simply two wings, flies are among the greatest aerialists in the world of insects as they can fly forwards and backwards, turn at any place, hover, and even fly upside and down to land on a top boundary. Flies have the uppermost wing-beat rate than any of other animal. It may be as high as 1000 beats per second in case of some small midges. Generally, through the wing-beat frequency of a virgin female, male mosquitoes are attracted. Maggots of certain shore flies (family Ephydridae) live in uncommon habitations, which would destroy other insects. For instance, *Ephydra brucei* survives in warm geysers and springs wherever the water hotness go beyond 112°F. The petroleum fly *Helaeomyia petrolei* Coquillett

Figure 1.
Wetland ecosystem.

develops in ponds of crude oil; and brine fly *Ephydra cinera* Jones, may live in extraordinary concentrations of salt [4].

The arista in the antenna of higher flies is an air speed indicator and it permits an insect to sense precisely just how fast it is moving. As black fly pupae mature, they become inflated with air. The pupal skin pops open upon emergence, and the fully-grown fly inside a bubble of air floats to water surface and it never even acquires its feet wet. The little scuttle fly *Megaselia scalaris* (Loew) is actually an omnivore. It has been cultured on paint emulsions, decaying vegetation, shoe polish, human dead body kept in formalin and even lung tissue from living people. As Diptera comprises the most ecologically diverse order of insects, swamps (wetlands ecosystem) resembling to this one (**Figure 1**) inundated by water and dominated by plant life, are great places to find its members.

3. Diversity among Diptera

Diptera have successfully colonized all continents including Antarctica, and are diverse not only in species richness, but also in their structure, habitat exploitation, life habits and interactions with humankind [5].

Maximum of nourishing and buildup of biomass take place in the larval stages and adults Diptera generally take energy they require to supply their flight muscles. Among those flies that forage widely, their foods contain honeydew or nectar (Blephariceridae and Bombyliidae), vertebrate blood (Culicidae and Glossinidae), pollen (Nemestrinidae and Syrphidae), insect hemolymph (certain Ceratopogonidae) and other biological resources that are liquefied or can be suspended or dissolved in regurgitated fluid or saliva (Muscidae, Calliphoridae and Micropezidae). The grownups of several groups are predaceous (Asilidae, Empididae and some Scathophagidae), while those of a few Diptera (Oestridae and Deuterophlebiidae) that totally lack of mouthparts, do not take food and live only for a short period [6].

Larvae of most species can be considered aquatic for existence, they need moist to wet atmosphere inside living tissues of plants, within decaying organic matters, as parasitoids or parasites of animals, or else are in link with water bodies. Maximum of larvae are free-living and crawl, tunnel or swim vigorously in water (Culicidae, Chironomidae, Chaoboridae, Simuliidae), sediments (Tabanidae, Tipulidae, Ceratopogonidae, Psychodidae), wood (Axymyiidae, Tipulidae, some Syrphidae), fruits (Tephritidae, Chloropidae), or decomposing biological material (Muscidae, Ephydridae, Sphaeroceridae, Sarcophagidae), whereas other larvae dwell in the tissues of alive creatures (Oestridae, Acroceridae, Tachinidae, Pipunculidae) [7].

4. Prominence of Diptera

The utmost essential significance of flies is not based only on limited acquainted families that comprise mosquitoes, tsetse flies, houseflies and other annoyance insects, but rather in the huge numbers of unfamiliar species that are a vital component in food chains upon which much of life rests on. Flies are of considerable ecological importance and their abundance, worldwide distribution and habits combine to make them a nuisance to humans by landing on people or entering homes or businesses. Midges and gnats are common names for a large number of small, non-biting flies and an important part of aquatic food chains. Swarms or clouds of midges in the air are a collective nuisance. Face flies and sweat flies, gather nearby the mouth, eyes and nose, and likewise suck pus and blood from sores and wounds. Such flies constantly move from one individual to the subsequent and in doing so, sometimes can transmit disease-causing pathogens [8].

The order of true flies contains more species with aquatic stages than any other insect group. Unlike all other invertebrate orders, dipterans contain many species that as adults are harmful or at least annoying to humans. At the top of the list are mosquitoes; however other harmful groups comprise horse flies, black flies, deer flies and biting midges. Various fully-grown dipterans transfer pathogens or parasitic diseases that can be fatal or devastating to persons, such as dengue, malaria, yellow fever, and West Nile virus. Furthermore, other flies (some midges) develop in such great numbers that they may cause allergic responses in persons or else block air conditioning items. Conversely, dipteran larvae are tremendously essential in aquatic food webs and specific groups are raised in hatcheries as fish diet, and several fully-grown dipterans are a vital food for dragonflies and birds such as swifts, swallows, flycatchers and phoebes [9].

In warm countries, eye gnats are an annoyance and although their larvae are plant feeders, the tiny active adults forage on physiological secretions, generally those around the eyes. Additionally, few flies cut the skin of vertebrates and nourish on their blood. Sand flies, mosquitoes, black flies, horse flies and biting midges have developed maxillae and blade like mandibles with piercing stylets. Such piercing organs are evolved only in females, which for egg production usage blood protein, whereas male dos not forage on blood [10].

Other flies groups have developed diverse devices for attaining blood meal. Stable flies, tsetse flies or biting house flies (*Stomoxys*), and some parasitic flies have evolved a tough drill like labium to substitute the soft sponge like mouth part. Both females and males have developed this labium and they forage on blood. A small number of flies correlated to the house fly have a spongy proboscis furnished with minor teeth for scratchy skin around sores and wounds to raise lymph and blood flow. Other insects (robber flies) have evolved piercing proboscis merely used against other insects.

Spread of disease that takes place by use of piercing organs such as a proboscis is reflected as mechanical transmission. In the blood, disease-producing organisms might be picked up by a fly introducing its proboscis into an infested individual. Then disease can be transferred by blood sucking fly, which injects its saliva into the wound of other persons when their skin is pierced. Without anticoagulant properties of saliva, blood sucking would be difficult as the minute hole drilled by proboscis would block with coagulated blood. When mouthparts are contaminated with blood that contains microorganisms, they can be injected together with saliva, into another person; this is termed as direct transmission of disease. One contagious disease caused by a bacterium found in wild rodents is tularemia that may be transferred in this way. Trappers who cut themselves while skinning animals can contract with the disease. The bacterium is also transmitted by deer fly (*Chrysops discalis*), Williston common in wooded trapper country.

In the Middle East and parts of Asia, surra is a disease of horses and camels caused by the protozoan *Trypanosoma evansi* and transmitted by horse flies. Trypanosomes are transferred by tsetse flies that cause sleeping sickness in humans and nagana in animals all over tropical Africa. These trypanosomes essentially pass a portion of their life cycle in the insect prior to they can contaminate a vertebrate host and this is an example of cyclic disease communication. The connection between two hosts, vertebrate and insect and parasitic disease organism is a result of evolutionary adaptation. On the other hand, it is not recognized whether the trypanosome originally has a fly parasite that speeded to humans and other vertebrates or else whether it has a human parasite that became adapted to live in a biting fly [11].

Malaria is a cyclically transmitted significant disease and causal mediator of human malaria, *Plasmodium*, is a cellular protist that feeds in human blood on red blood cells. Its propagative cycles cause repeated attacks of illness. Sometimes sexual forms take place in blood of victims. When this form catches its mode into an appropriate blood sucking mosquito, a different phase of *Plasmodium* arises, making an organism to contaminate one more human host bitten by mosquito species. Further illnesses recognized to be transmitted cyclically comprise encephalitis, filariasis, yellow fever and other viral infections [12].

Some flies larvae are severe pests of agriculture, they forage on young and mature crop plants and check growth otherwise destroy them. Cultured crops, for the reason that they offer to pests with a nearly limitless nutrition resource inside a small space, may be destroyed by uncontrolled density of a pest. In contrast, wild food plants, for the reason that they are mixed and scattered with other varieties, do not generally offer much plentiful of food supply and therefore work as a check on population growth. Fruit flies may result a 20% damage of an oat crop and to the value of the lost oats might be added the price of control actions essential to protect the leftovers. Several crops, particularly ornamental shrubs and fruit trees are of an economic injury if to some extent spoiled by insect invasion, although the life of plant is not threatened. Fruit, although is eatable afterward injured by Mediterranean fruit flies, yet cannot be traded since a limited infected fruits can result in loss of an entire consignment. Larvae of leaf miners and gall midges reduce the saleable price of ornamental plants [8, 13, 14].

Brown white-tipped brown bee fly *Comptosia walkeri* Edwards (brown in color with golden hairs on body and wings having white stripe on tips) (**Figure 2**) and fly *Comptosia insignis* (Walker) are members of family Bombyliidae. These flies superficially resemble to bees owing to their bodies built stoutly that are shielded with hair and by bearing their long and thin proboscis. These together with their flight lifestyles have gotten them collective name of bee flies or humble flies. Adults may be seen frequently hovering over or resting on blossoms or areas of plain ground in sunlit localities. Adults of bee flies suckle on nectar from a widespread diversity of floras and can be key pollinators of plants. Even though a slight is recognized about

Figure 2.
Comptosia walkeri.

some species, bee flies larvae are thought to parasitize the larvae of other insects and may prey on egg-masses of locusts and grasshoppers [15].

Family Tephritidae comprises peach fruit fly *Bactrocera zonata* (Saunders) mainly pest on peach and other stone fruits; guava fruit fly *Bactrocera correcta* (Bezzi) common on guava, mango and citrus; oriental fruit fly *Bactrocera dorsalis* Hendel recorded on fruits and vegetables; and melon fly *Bactrocera cucurbitae* (Coquillett) that is a major nuisance of cucurbitaceous vegetables (**Figure 3**) [16–19]. Overall, the damage to fruits and vegetables caused by fruit flies results from adults oviposition in hosts and soft tissues of vegetative parts of certain plants, feeding by white legless maggots, and decomposition of hosts tissue by invading secondary microorganisms [20–23]. Integrated pest management (IPM) packages for fruit flies include, mechanical controlling by protective coverings on the host and the destruction of adults with baiting and male annihilation techniques, biological control with biopesticides and parasitoids, cultural control through field sanitation and resistant varieties, using selective pesticides for backup only, and post-harvest control by careful host selection and hot water immersion treatment [24–38].

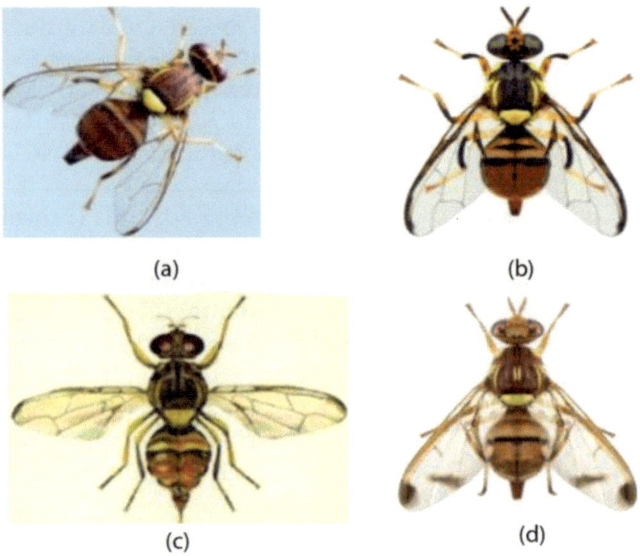

Figure 3.
(a) Bactrocera zonata, *(b)* Bactrocera dorsalis, *(c)* Bactrocera correcta, *(d)* Bactrocera cucurbitae.

Figure 4.
Musca domestica.

DOI: http://dx.doi.org/10.5772/intechopen.91391

The house fly (*Musca domestica* Linnaeus) (**Figure 4**) can be dangerous because it moves from person to food, drinks, garbage, carrion or feces. It is possibly our most public adult fly and though this is a non-biting fly, it may be of significant prominence as a nuisance. In addition, it has a probable role for mechanical transmission of several sickness causing agents owing to its occurrence in fecal and decomposing organic matters. Through transferring of infective organisms from decomposing material or from infected people, house flies are agents in transmitting of typhoid, dysentery, cholera, summer diarrhea in children, and other intestinal virus and bacteria causing diseases.

The house fly is about 6–7 mm in length and females generally bigger than the males. The female may be differentiated from the male by means of comparatively broader space among the eyes (in males eyes mostly touch). The head of adult fly has reddish eyes; thorax bears four narrow black stripes, while abdomen is gray or yellowish with dark midline and irregular dark markings on the sides. The mouth parts of this fly are modified for lapping up of food material. The egg is about 1.2 mm in length, white colored and singly laid; however, eggs are stacked in small clusters. Every female fly may deposit up to 500 eggs in a number of groups containing 75–150 eggs in a period of 3–4 days. The legless maggot emerges from the egg in warm weather within eight to 20 hours. Early instars larvae are 3–9 mm long, typical creamy whitish in color, cylindrical, but tapering toward the head. The larva goes through three instars; a full-grown maggot is 7–12 mm long and has a greasy, cream-colored appearance. The pupal stage, about 8 mm long, is passed in a pupal case formed from the last larval skin which varies in color from yellow, red, brown, to black as the pupa ages. Pupae at 32–37°C, acquire their complete development in 2–6 days, but at about 14°C need 17–27 days. The evolving fly discharges from pupal case by the usage of a consecutively shrinking and swelling sac known as ptilinum, to breakdown through the case.

Generally, warm season situations are ideal for the development of house fly and in as little as 7–10 days, it can complete its life cycle. On the other hand, the life cycle may require up to 2 months under suboptimal conditions. In temperate regions, as many as 10–12 generations may occur annually, whereas in subtropical and tropical regions more than 20 generations can take place [39].

Face fly *Musca autumnalis* DeGeer is a serious pest of cattle though this is a non-biting fly. Adults come together in great numbers around the face of cattle and other big ungulates. This has been involved in the spread of 'pink-eye' between cattle and its occurrence in great numbers has been liable for indirect harms in production comprising reduced weight gains and milk production. The adults are at maximum vigorous from spring through fall.

5. Anatomy and physiology

Any member of the order Diptera has evolved a simplified structure and physiological diversity. Dipteran larvae can be differentiated from maximum of other insects by means of their absence of segmented thoracic legs. As a replacement for the customary jointed legs, several crowds have one or more couples of fleshy locomotory prolegs on abdomen and or thorax, each with curled or even hook-like spines. Fleshy tubercles arise in several species and help together in locomotory and sensory tasks. The larva head may be heavily sclerotized and exposed, as in midges, or else toughly condensed and only moderately proud (from time to time only with mouthparts expanded). The abdomen and thorax are generally fleshy, from time to time with dispersed sclerotized plates, and the whole body is usually long and tubular, mean lengths are 2–25 mm, however can range 10 cm in several species. Always, wing pads are lacking in larvae, however existing in pupae.

Various larvae breathe through the skin and minute gills exist in certain taxa. Others dipterans get oxygen using spiracles and either lengthy or small breathing tubes from the atmosphere (as in mosquitoes). Limited groups remove oxygen from plant tissues. Certain true midges like blood worms (a type of true midge), which to some extent frequent anoxic habitations, have an invertebrate form of breathing pigment hemoglobin that supports in catching of oxygen molecules.

Adult dipterans vary in size from 1 to 12 mm; however, comparative giants of 25–60 mm long are identified wherein the later contain bigger crane flies. They have a single pair of membranous wings; hind wings are rudimentary and nonfunctional for flying whereas the bodies are long and tubular. In feeding adults, the mouthparts are modified to sharp tubes for penetrating flesh and sucking up liquids as in mosquitoes or adapted for consuming liquid food using either blunt pads for sponging up liquid. Although crane flies are from time to time called 'mosquito hawks', these do not consume mosquitoes, nor they bite human being.

6. Reproduction and life history

Entirely, Diptera have a complete metamorphosis (holometabolous), meaning that they go through four life stages, for instance, egg, larva, pupa and adult. The number of eggs laid by a female varies by species, from just a few eggs to thousands of them. Mom does not have any involvement in the care of its babies, so it lays eggs on a food supply where they hatch. Females place their eggs in clumps or singly, generally near water and from time to time attached to other things. Eggs have a tendency to last only for limited days with the exception of diapause eggs, which are used to avoid unfriendly temperatures or the shortage of water in environment. Larvae, which often look like worms, hatch from the eggs and after hatching, the larvae of maximum species traverse three to four instars (six to seven in black flies) prior to pupate on land or near bottom or at water surface. The larval stage lasts for somewhere from nearly 2 weeks to some months. These larvae may have 'false legs' called prolegs that look like the little legs seen on caterpillars. But, Diptera larvae lack any truly jointed legs. As larvae are always divergent morphologically from adults and moreover live in different habitats, flies basically spend two distinct lives and thus are capable to adjust environmental changes successfully. In several flies (robber flies), neither larval nor adult stage predominates, their larvae actively forage in soil and both sexes of adult flies in flight catch other insects. Among mosquitoes, black flies and correlated blood sucking flies, larvae have distinguishing structures and spend active lives under water, and the complex mating method of adults is followed by blood sucking and egg laying (in the case of females).

There are several flies in which one stage is predominant, for example, groups of adult midges (Chironomidae), are noticeable and bothersome, however adults midge live just long enough generally fewer than a day to mate and lay eggs. The maximum of life cycle by the larval stage is occupied under water. Larvae in appearance are wormlike and certain are adjusted to oxygen-poor conditions, for example, the 'blood worm' that lives in the sludge of standing waters, usages hemoglobin as a breathing pigment. Larvae of few midges live in silken tubes, either filtering minute organisms from water for food or preying upon other creatures. Certain midge larvae have developed an elaborate mutualism or symbiosis, using other aquatic creatures, for instance, certain midge larvae and *Nostoc* (a genus of blue-green algae) utilize excreta of each other.

At the opposite extreme are tsetse flies (*Glossina*) and three families of pupipara (females birth young ones) parasites (Nycteribiidae and Streblidae feed only on bats blood, and Hippoboscidae feed on the blood of mammals and birds). Within these families, single egg is formed at one time and it hatches internally. The larva is nourished and retained within a kind of womb, ejected out after this has matured and instantly forms a pupa. Hence, larvae of these flies have not an independent life. As pupa is immovable, the active life of fly is spent as an adult. Maximum of Streblidae and Hippoboscidae, and whole tsetse flies, bear wings and generally transfer to different hosts, however certain species of these families and entire Nycteribiidae, cannot fly and are frequently wingless. Wingless flies might be recognized as flies afterwards thorough morphological checkup only. The lifespan of an adult varies by species and may extend from a few hours to weeks or months. Even though a multivoltine life cycle is mostly prevailing, certain dipterans create one generation only in each year (univoltine) and might need many years to create a fresh generation in arctic and further cool atmospheres. The shorter generation periods illustrating this order allow dipterans to react rapidly to fresh diet or other resources and to recover quickly from critical circumstances like stream waves.

6.1 Adults

There are many different shapes of true flies and they are soft-bodied insects, most are fairly small (less than 1.5 cm long), but a few can be larger (up to 4 cm). The adult body colorations of different fruit fly species vary from black through various shades of brown to orange or yellow. **Adult** antennae are filiform, stylate or aristate, mouthparts suctorial (haustellate), mesothorax larger than pro or meta-thorax, one pair of wings (front), hind wings reduced (halteres) and tarsi 5-segmented. Adult fly appears from soft pupa and is lined with a colorless integument (skin), and have made perfectly (although not fully pigmented) bristles and hairs. The newly evolving adult sips air to swell its wings and body, and to power flow of blood through body. In more progressive flies of set Schizophora, in head, an expandable membranous sac ptilinum is used to help this procedure. The ptilinum shrinks away afterward it has done its job, but behind, it leaves ptilinal suture, which is a horseshoe-shaped furrow growing over and alongside antennal sockets and found in Schizophora only.

6.2 Eggs

Mostly, flies lay eggs that hatch into minute larvae afterward a few hours or some days. The eggs number put down by one female fluctuates from 1 to around 250. But, a large number of succeeding eggs batches can be laid. Green bottle fly (*Lucilia sericata* Meigen), in confinement has set almost 2000 eggs. But, the number total is perhaps less than 1000 in ordinary environment when energy and time are gone seeing for proper places for laying eggs. Sites for egg-laying, preferred by females instinctively, are closely linked to habitats of larvae. As many larvae of fly forage into soft organic maters, several females have established telescopic ovipositors, made from either last three otherwise four segments of abdomen. The female uses an ovipositor to press its eggs within a mass of rotting matter. House flies and blow flies thrust their eggs among membranes of meat otherwise into any suitable cavity in rotting organic matter. Small fruit fly (*Drosophila* vinegar fly) that lays eggs in decomposing fruits and fermenting matters as well have this kind of ovipositor, but, large fruit fly (Mediterranean fruit fly) that lays eggs in the rind of developing fruits, bears a harder ovipositor. An elaborate ovipositors set up in robber fly is used to thrust eggs into axils of grasses and interstices of flower heads, and from time to time even into plant tissues, to protect and conceal them from drying. Once hatched, the larvae drop to the ground and burrow below the soil.

6.3 Larvae

The larvae of true flies look like thick segmented worms, but they have many different shapes. They do not have jointed legs, unlike beetle larvae. Some larvae have mouthparts and a distinct head, but most do not bear typical structures. Larvae of fly have single joint characteristic, wherein all have lack of jointed true thoracic legs. Several larvae of flies have 'false legs' (pseudopods or prolegs) related to those that care fleshy abdomen of caterpillars. Flies are greatly more adjustable than caterpillars in this respect, and around anybody segment can have prolegs. The prolegs aid larvae to push through soil or crawl into narrow spaces.

The evolutionary tendency among larvae of fly has been in the direction of structural simplification, so, usually, primitive flies larvae are extra structured than larvae of further importantly evolved flies that display better physiological adaptability. Most members of suborder Nematocera or Brachycera larvae bear a well-built head having antennae, complex mouthparts and palpi related to several adult insects. Frequently they are so structurally modified to their distinctive mode of life that these are not capable to adjust any other. It is particularly true among aquatic larvae such as mosquitoes and possibly reaches at extreme in larvae of mountain midge that live in roaring torrents and creep on immersed rocks. Segments of their body are furnished with suckers and clinging processes. The maggots of Cyclorrhapha have tiny external structure other than the posterior spiracles and black mouth hooks.

In comparison to very specialized larvae, nearly half species of flies have larvae called maggots. The maggots have missing the complex head capsule of primitive flies. Their sharp anterior ends comprise one or a couple of mouth hooks. The rounded posterior end has one couple of spiracles posteriorly (outer air holes), which look as black spots to the naked eye. Microscopically, spiracles are recognized as pores or a complex arrangement of slits, which are valuable in differentiating of species.

Even though maggots display structural homogeneity, they are dissimilar physiologically. Maximum of maggots feed upon rotting organic material, however in forensic studies; there are wide dissimilarities in the food likings of various flies. Larvae of gout fly of barley and frit fly of oat are maggots of flies that fit into plant-feeding family Chloropidae. The hessian fly of wheat is the destructive larva of *Mayetiola destructor* (Say) (gall midges) in nematoceran family Cecidomyiidae. Even though external structure of maximum nematoceran larvae is diverse, the structure of gall midges that live totally submerged in plant tissue, has developed in the direction of simplification. Gall midges fly larvae are known as well gall gnats for the reason that feeding larvae result in the development of disfiguring galls on stems or leaves and harm to several kinds of plants. So, they have developed physiological diversity and simplified structure concerning food floras as have maggots of further progressive flies.

The greatest well-known blow flies existing are sheep blow flies, essentially species in *Lucilia* genus. Maggots of *L. sericata*, forage on tiny deceased animals, and in garbage and abattoirs cans, oviposit in dirty wool around anus of sheep otherwise in pus oozing from wounds and scratches, wherein these are essential means of sheep strike illness. These maggots from time to time occur in soil adjacent to buildings in towns and their diet basis is not recognized. Eight 'waves' of worms have been well-known, and each wave assaults deceased animals in a severe series as decay develops from newly deceased carcass through thoroughness and decomposition to mummification. Though several maggots only appear during a noticeably definite stage of animal decomposition, the greedy large maggots of various blow flies nourish on any animal material, comprising living tissues [40].

The number of instars or larval stages is six or seven in black flies (Simuliidae) and four in most other Nematocera. Alongside the second line of evolution of flies, Brachycera have from five to eight instars, while Cyclorrhapha maggots of the most advanced flies have only three instars. Three economically important free-living instars exist for tephritid fruit flies. The *Urophora jaceana* (Hering) and *Urophora cardui* (Linnaeus) have the first instar remaining in the egg and exits as a second instar. One or two species have no molts and from time to time molts arise earlier to the larva hatches from the egg. Muscidae, for instance, are organized in three groups according to whether they are monomorphic (pass the first two instars in the egg, have one free larval instar), dimorphic (pass the first instar in egg, have two free larval instars) or trimorphic (have three free larval instars). Monomorphic larvae are constantly predatory; however, dimorphic and trimorphic larvae feed first on decaying matter (saprophagous) and, on the other hand, they may or may not be predatory in their final instar [41].

Larval breathing is adjusted to the medium wherein larvae live. Even though limited parasitic larvae (Pipunculidae, parasite in Drosophilidae and froghoppers, internal parasite of scale insect) take oxygen by the skin and maximum dipterous larvae require tracheal system to allocate oxygen. Basically, tracheal system possibly is exteriorly opened on each body segment by paired spiracles. The soil occupiers, Scatopsidae and Bibionidae, hold this system, even though maximum families have retained spiracles only on thorax (one pair) and one at the abdomen tip. These are even sealed in several aquatic larvae (larvae of biting midges and luminous larvae of some fungus gnats). On the other hand, larvae of mosquitoes and many other water-living fly larvae repeatedly come to surface for renovation of their oxygen provisions. Certain larvae of flies pierce twigs of underwater plants to get oxygen made by photosynthesis activity. In Cyclorrhapha, maggots heavily depend on posterior spiracles complex. Pupae take breaths by prothoracic spiracles, which are from time to time furnished with long tubes extending outside the puparium or cocoon.

6.4 Pupae

The pupal stage of a true fly is enclosed within a hard capsule (skin). It may have some of its legs and body parts visible or it may be hidden inside a larval skin and just looks like a brown capsule. Dipteran pupae have non-functional mandibles (adecticous) and may have the appendages free from the body (exarate) or glued to the body (obtect). In exarate type, the pupa is concealed inside the hardened skin (puparium) of the last larval instar. The external structures of adult fly (antennae, eyes, legs, wings) are obviously noticeable in the pupa. However, the pupa, is not every time visible to the sight, it may be encircled either in a puparium that is a case formed by toughening of the larval skin or in a cocoon of extraneous matter (silk, soil or a mixture of the two). In flies of Stratiomyidae family and others, which have maggots like larvae (whole Cyclorrhapha), a puparium is formed. Many fly families sporadically form cocoons and cocoon has developed an adaptive tool, which delivers an extra safety to the pupa. Pupae of mosquitoes, black flies (Simuliidae) and limited aquatic midges swim vigorously. Several pupae that lie in wood or in soil have evolved spines in order to aid them for effort to have their way to the surface just before appearance of adult insects.

6.5 Wings

Adult flies are usually active during the day when it is warmer and they also sometimes detect the vibrations of wing beats. Adult flies have only one pair of wings on the mesothorax or second thoracic segment. The hind wings, modified into small club-like halteres behind the much larger forewings, have a knob or club and a stalk, which may be big and thick comparative to the size of fly. The halteres vibrate above and below in time with wings and in flight perform as gyroscopes (maintain or measure motion). If fly rolls, yaws or pitches in the course of flight, halteres maintain their original plane of movement, twist at their roots, where special nerve cells identify the twist and cause fly to accurate its flight attitude. The base of halteres is elastic and when these are stirred, a fly is capable to control its flying. As the halteres curve at the base, a fly can change flight direction or speed thus making them well controllable in comparison to various other flying insects.

The wings of flies have a well-defined pattern of veins; each has a characteristic location and name, and often has taxonomic significance. A small number of true flies have a reticulation (network of small veins), nearly resembling to various other insects, which are called flies (dragonflies, mayflies, dobsonflies) mistakenly. Primitive flies tend to have complex wing venation, whereas advanced ones have simplified and reduced venation. Some of the small midges (Sciaridae, Cecidomyiidae, Mycetophilidae) have also reduced wing venation. Reduction or losses of wings take place in several families, predominantly, which dwell windy dwellings (islands, mountains) or caves, or those are exterior parasites among feathers and furs.

6.6 Eyes

Most adult flies have large eyes, to help them see when they are flying. Flies use vision more than most insects do and like all insects, they use their sense of smell a lot. The eyes of most flies often lodge on much surface area of head, particularly in males, where eyes may well come across in the middle line (holoptic). With few exceptions, in female flies, the eyes do not normally meet (dichoptic). In certain families, notably both sexes of small acalyptrate flies and robber flies are dichoptic. Parasitic flies or those that living in sheltered dwellings can have very little eyes or none of any kind. Characteristically, on the other hand, compound eyes of flies comprise several facets, for instance, house fly in each eye has 4000 facets and some *Drosophila* sp., has 700 facets per eye that help them to see.

6.7 Mouthparts

Characteristically, flies have suctorial type of mouthparts and several bear large fleshy pads along with drainage canals called pseudotracheae for proficient uptake of liquid. Mouthparts of certain flies are modified for piercing and stabbing of other insects, for instance, predatory dance flies (Empididae) and robber flies (Asilidae). Mosquitoes and certain other ectoparasitic insects have modified mouthparts for penetrating into vertebrate host skin, and take out blood and other body fluids. In various families, rostrum (proboscis) is altered for lapping and or sponging. These flies live on nectar, honeydew or exudates of different plants and animals (alive or dead). In further families, proboscis is amended for piercing or cutting the tissues of hosts. Many of these flies are outer parasites (mosquitoes and deer flies) that feed on the blood of their vertebrate hosts, including humans, and most wild and domestic animals [42].

Many flies have maxillae, most have also mandibles and stretched blades that cover a furrow in labium and arranged as tubular channel for sucking liquids. In many females (mosquitoes, blood sucking flies), for drawing blood, mandibles act as piercing stylets. Mandibles have been lost relatively entirely early in fly evolution or became functionless and as a result families of blood sucking insects, which afterward evolved have to develop other methods of piercing the hosts. Stable flies and tsetse flies usage toughened labium, dance flies and robber flies practice hypopharynx, and Dolichopodidae (metal green flies having very large legs) crush their prey with especially evolved teeth by wrapping in spongy labella of labium. Many flies suck their diet, whereas with few exceptions have condensed mouthparts and probably do not forage at all as adults. Therefore, diet of flies might be liquid, otherwise solids, which may be dissolved by stomach juices and saliva. Flies as well bear a couple of labial palpi fitted with sensory cells, which act as organs to detect smell, taste and touch. The antennae and palpi are important for scrutinizing of probable diet sources and appropriate spots for laying eggs.

6.8 Antennae

Entirely, flies bear antennae and great antennal structural differences occur among related species and genera. Members of suborder Nematocera (crane flies, midges and gnats) have whip-like antennae with two basal segments (scape and pedicel) and apical flagellum of many similar segments. Altogether in other flies, accurately called Brachycera, flagellums are contracted into a compound third segment and have remnants of the terminal flagellar segments remaining as a bristle-like arista or a pencil-like style.

7. Ecology of Diptera

Diptera are such a diverse group that they can be found just about anywhere and these are most common in humid or moist environments, but can also be found in deserts, forests, mountains and even polar regions. They are also common in both fresh and saltwater environments such as lakes, ponds, streams, marshes and swamps. There is hardly any life-supporting medium in which dipterous larvae have not been observed. Species of Diptera can be gathered in wide range of habitats from most polluted to most pristine environments, from fast flowing water to stagnant water or from saline water to freshwater. Pupae and larvae are found among aquatic vegetation, organic debris, problematic habitats, sand, fine sediments, gravel, mud, cobbles or bedrock. They might be restricted to and sometimes closely associated with water surface, water column, any of aquatic zones, main water flow, benthic, littoral or interstitial and hygropetric zones. However, maggots are the utmost essential larvae, for the reason that they perform a crucial part in restructuring and breaking down organic material. The waste produces expelled by the larvae offer nutrients for molds, fungi and other types of plants. Additionally, bodies of larvae, pupae and a lot of adult flies are essential diet sources for higher animals. Cases in point are aquatic larvae of mosquitoes and midges that are basic diet for fish. Also, the terrestrial maggots of various flies have a part in nutrition chains. Meanwhile, a blow fly is able to lay one to two thousand eggs; their density would upsurge terribly if more than a few of them stay alive. Maximum of the larvae pass away owing to desiccation, malnutrition and sinking or are used up by birds. Adult flies are snapped up by small mammals, birds, toads and frogs. Martins, swifts and swallows consume huge numbers of flies that have been brought into the air by convection currents. So, their density is conserved at a persistent level.

Within more primitive families in suborder Nematocera, larvae of flies have well-built head capsules having mouthparts of mandibulate type. These arrangements are absent or reduced in more progressive Brachycera and Cyclorrhapha suborders wherein larvae are recognized as maggots, having worm-like bodies and a couple of mouth hooks only for nourishing. The abdomen, thorax and legs of adult flies differ from short to long and appearance of fly is well-designed along with decorative style. From time to time, bright color and pattern of several flies (blow flies) is metallic, on the other hand, most often fly is concealed with a good coating called dusting or tomentum. Numerous flies, principally those of more greatly evolved families, are bristly and the strongest bristles have an accurate location, mostly on thorax. The identification of bristles, their arrangement and the method established on them is known as chaetotaxy [43, 44].

8. Nutritional requirements of Diptera

Nutrition involves balance between feeding habits of larval and adult flies, and primary feeding occurs during the larval stage. Adult feeding serves to compensate the shortcomings of larval nourishment. Adult flies often drink upon fluids, but some feed on any liquid that has nutrients. They also can 'spit' onto dry food and then suck up the spit and some extra nourishment from the dry food, and thus infect human food. Certain female flies suck vertebrates blood, for instance from mammals to get protein they need for their eggs. A small number of adults are predators; they grasp other insects, crack them with their mouthparts and draw out their organs and fluids [45].

Generally pupae do not feed, and several flies do maximum of their nourishing as larvae and fly larvae often feed continuously during day and night. Certain feed on plants or eat fungi, but mainly upon fruits. Certain species are gathering collectors (feed on organic detritus), filtering collectors (feed on suspended diatoms and fine detritus filtered from water column), scrapers (use mandibles to scrape algae and fungi), shredders by chewing and boring (feed on leaf litter or living macrophytes) and predators prey on other invertebrates including their own species. Some place their eggs in leaves or stems and their larvae emit chemicals that make the plant to swell up into a gall. This defends the fly larva and provides to it a sufficient of nutrition to feed. Further species consume deceased animals and several feed on dung. Certain filter microscopic diet elements from freshwater (river, lake, stream). A single large group of flies in nature is parasitic, these deposit their eggs outside or inside of other insects and spiders, and their larvae nourish on inside of hosts even though host is still thriving. A limited species are vertebrate parasites, like birds and mammals and live under the skin or into wounds of their hosts. Several dipteran larvae live in aquatic, semi-aquatic or wet terrestrial atmospheres. They are normally found in soil, animal tissues or plant and in carrion or dung, where there is almost always a little risk of desiccation. Certain species are herbivores; however maximum forage on dead biological matter or parasitize other animals, particularly vertebrates, mollusks and many other arthropods.

At one extreme are nonbiting midges, with larvae that vigorously filter microorganisms from water. Correlating to nonbiting midges are biting midges, black flies and mosquitoes. Female adults in these families need supplementing diet in an insufficient larval food. Even though one set of eggs rarely is put deprived of a blood food, but blood is essential to develop a subsequent lot. Flies, which place one egg batch devoid of blood, are termed autogenous and those that cannot lay without blood at all are anautogenous. One species may have both types, probably as a result of unstable populations otherwise races rising from usual selection. Such as far north, great densities of biting flies (horse flies, black flies, biting midges mosquitoes) arise for the period of small Arctic summer and there are noticeably inadequate amounts of warm-blooded animals to offer diet. If flies clip blood, they consume it; however, these still stay alive, if not availed.

The adults do not feed and most flies visit flowers, which provide water, nectar and pollen. Although the name *Drosophila* means 'lover of dew', this insect sucks water and any other obtainable fluid. Nectar from flowers contains carbohydrates and most adult flies use this syrupy liquid. Pollens are tougher to obtain for a sucking insect than blood, which is rich in protein and a vital source of this nutrient. Several hover flies love pollen grains among hardened portions of labella prior to swallow them and certain flies actively probe into flowers, covering their eyes and heads with pollen grains. Although their role in pollination is less well known than that of bees, flies are important pollinators of flowers. Some plants (spurges) are often covered with small flies of different families. Small flies also feed on honeydew from aphids or whiteflies. Flies forage on dung and fluid produces of either vegetable decay or animal. They get nutrients from garbage dumps and farmyard manure heaps. These dwellings as well harbor several larvae, which nourish either right on biological diet available or feed on other larvae as carnivorous. Yellow dung fly larvae and adults, is a familiar example that target on other insects coming to dung [35, 36, 46].

Adaptableness of flies is obvious on a widespread range of foods consumed by the larvae. Aside from parasites, the maximum specific feeders are those larvae, which live in plant tissues (leaf mining Agromyzidae, may be limited to group of plants or one plant species). Commonly, pests of horticulture and agriculture (cabbage root fly) are multipurpose species, nourishing on a diversity of wild type hosts and altering their foods while offered with intense plantings of marketable crops. Numerous carnivorous larvae of fly (asilids) most likely reside in soil and consume animal or vegetable material, whatsoever is accessible. Meanwhile adult robber flies (asilids), forage on various insects, their larval diet is recognized to be insufficient. Certain maggots, predominantly young insect, which forage on plant material for the duration of second and first instars, turn into carnivorous in the course of third instar, where maximum of development occurs.

Adult flies escape from predators with their alertness and speed. Likewise, several flies mimic stinging insects, for instance, bees or wasps, therefore predators will avoid them. Larvae habitually live in dwellings that are hard to reach by predators. Well-known predators of flies are shrews (eat larvae and pupae), rodents (pupae), moles (larvae and pupae), toads (mostly adult flies), frogs (mostly adult flies), birds, ants, wasps, other flies, spiders, ground beetles (larvae and pupae) and true bugs (larvae and pupae).

In many cases, only the adult females of biting flies under certain circumstances get blood diet, Culicidae family of mosquitoes, possibly spreads dengue, malaria, filariasis, encephalitis, yellow fever and other illnesses. Tabanidae (deer flies/ horse flies) can transmit loiasis, trypanosomiasis, tularemia and some other sicknesses. Simuliidae family of black flies feasibly spread onchocerciasis of human and leucocytozoon contaminations in poultry. Moth flies of Psychodidae can transmit leishmaniasis, sand fly fever and further diseases. Family Ceratopogonidae having punkies and no-see-ums are small and on the other hand vicious biters associated to transmit some protozoan, roundworms and virus pathogens in animals and humans. Muscidae family of house flies is among the utmost cosmopolitan than all other insects. Particular species have piercing mouthparts while some others are only scavengers, and diseases such as cholera, yaws and dysentery can be transferred on their mouthparts and feet.

Larvae of herbivores flies forage on plant tissues, some gall midges (Cecidomyiidae) bring creation of plant galls, some are parasites, predators or scavengers and this family primarily comprises the Hessian fly *Mayetiola destructor* Say. Numerous species of family Tephritidae (fruit flies) are agricultural pests, such as the apple maggot *Rhagoletis pomonella* (Walsh). Most larvae of family Agromyzidae are leaf miners, some are stem and seed borers, and many Anthomyiidae species are seed otherwise root maggots.

Scavenger larvae nourish in garbage, carrion, dung otherwise further biological material, pomace flies (Drosophilidae) larvae forage on rotting fruit, crane flies (Tipulidae) live in mud or soil, larvae of blow flies (Calliphoridae) nourish on carrion or garbage and include screwworm *Cochliomyia hominivorax* (Coquerel), midges (Chironomidae) aquatic larvae generally breath in mud and forage on biological material, and larvae of flesh flies (Sarcophagidae) usually nourish on carrion, but certain species can be source of human myiasis.

Predatory larvae or adults predate on other insects as prey and specific flies produce predatory maggots that feed on other maggots. The predatory maggots of *Chrysomya rufifacies* (Macquart) (bluebottle) are covered with spiny protrusions which deter other predators. Families such as Asilidae (robber flies) are wide-ranging predators of other insects, Bombyliidae (bee flies) has predatory larvae and adults mimic to bees, Syrphidae (flower flies) (**Figure 5**) have larvae that are aphid predators and maximum adults mimic wasps or bees, and Empididae (dance flies) adults are also predatory. Parasitic larvae are parasitoids or parasites of insects and animals. Families like Tachinidae have several species that are important biocontrol agents and parasitoids of other insects, Hippoboscidae (louse flies) adults are blood-feeding ectoparasites of mammals and birds, larvae of Sciomyzidae (marsh flies) parasitize snails and slugs, and Oestridae (warble flies and bot flies) larvae are endoparasites of mammals as well as humans [47].

Figure 5.
(a) Robber fly, (b) Dance fly, (c) Flower fly, (d) Louse fly.

9. Roles in the ecosystem

The prime advantage of flies originates from the parasitic species. They invade grasshoppers, caterpillars and other insects that damage to food plants. Certain flies are also important pollinators and aid to pollinate plants that are grown. Flies are likewise essential food basis for other animals that are valuable such as fish. Many fly larvae are part of the natural clean-up squad, helping to get rid of dung and dead animals.

9.1 Decomposition: fly life cycle and development times

Marine Diptera, which perform decomposers function are land-dwelling muscoid flies, for instance, in aquatic habitations, Calliphoridae are under-represented mainly when a carcass is fully underwater. Corpses floating on surface of water and alongside shorelines, deal a terrestrial-aquatic boundary, which is expected to comprise dipteran agents of both surroundings. Certain ephydrids of shore-inhabiting devour minor carcasses of animals on seashores, suggestive of their legal prospective to as well practice bigger carcasses. Forensic circumstances in water linking to a deceased person normally feature presence of chironomids, their abundance, diversity, ubiquity and species-specific appearances, which make to Chironomidae possibly beneficial in criminal inquiries. Adhoc examples to utilizes marine Diptera in legal inquiries have been recognized. Exuviae of black fly *Prosimulium fuscum* (Syme & Davies) (**Figure 6**) pupae found on a submerged case helped to convict a murderer by countering his purported timeline of events [48].

Forensic entomology creates use of information resulting from either the series of arthropods on animal carcasses or human corpses or temperature-dependent development of insects (principally flies) to appraise the time gone since passing away or postmortem interval (PMI) or estimate of the time between death and corpse detection. Flies are the most significant organisms for forensic study and especially valuable in determining the age of corpse from duration of a few hours to a few years [49].

The occurrence of insects within any carcass is a serious sign toward guessing death time of bodies deceased for lengthier time periods. Since, flies quickly find out a body and their times of growth are foreseeable during specific ecological circumstances, so death time may be determined by calculating back days from the state of fly's growth existing on carcass.

Figure 6.
Black fly.

Fly species	Egg	First instar	Second instar	Third instar	Pre-pupa	Pupa	Total time (days)
Hydrotaea rostrata	48	60	60	36	144	324	28
Chrysomya rufifacies	24	36	36	72	72	168	17
Calliphora augur	no eggs	24	24	60	96	336	23
Calliphora stygia	24	48	24	48	96	324	23
Lucilia cuprina	26	33	33	24	114	324	23
Lucilia sericata	21	31	26	50	118	240	20

Table 1.
An average approximate development times (in hours) of some specific fly species at 20°C.

Blow flies (family Calliphoridae) are metallic blue, green or black in color, noisy in flight and resemble to the housefly. A female of blowfly at one time lays up to 300 eggs and with many females visiting any corpse, number of maggots may be immense. For instance, on a 156 g piece of meat, 48,562 maggots are initiated after 24 hours exposure. On the other hand, since this has been inadequate food to withstand them, only 231 flies lastly emerged. In hot weather, helpful to fly growth, maggots can devour 60% of a human body in less than a week.

The growth times of fly vary depending on the species and the temperature, but generalized life cycle typically takes 3–4 weeks depending on the species. Eggs are found in clusters of up to 300 and time takes 1 day from laying to hatching. Initially, first instar larva feeds on fluid oozed from body then migrates into body and takes 1 day from hatching to first molt. Larva of second instar travels around in maggot mass and from first molt to second molt takes 1 day. Larva in third instar moves still in mass, significantly increases in dimension and takes 2 days from second molt to pre-pupa. Pre-pupa drifts away from the corpse for seeking an appropriate pupation location (commonly in soil), does not forage, converts into pupa and takes 4 days from pre-pupa to pupa. Pupa exist within puparium, does not feed and transformation from pupa to adult fly emergence takes 10 days. Upon emergence from pupa, adults fly mate, feed on protein from body fluids, and lay eggs on corpse and emergence to egg laying takes 2 days [50, 51]. The life table (**Table 1**), shows an estimated development times (in hours) at 20°C of certain fly species.

10. Phenology

A characteristic life-cycle of dipteran follows a short-term egg stage (generally days or on occasion greatly longer), larval and pupal phases of variable length, and an adult stage lasting for a few to many hours or days. The period of larvula is shortest, while the last larval stage that is key feeding form is much lengthier. Totally, larval instars share an identical habitat, however various Chironomidae make sure to have planktonic larvulae and benthic later-instars. A lot of marine Diptera is univoltine categorized by quick development. In cyclical system with cold season, immature insects generally in an initial instar, diapause till environments are satisfactory. Postdiapause development regularly initiates with increasing spring hotness, even though algal accessibility and photoperiod may be associated. Period after egg-hatch to adult beginning differs among and occasionally contained by species, such as does presence of further generations (bivoltine to multivoltine).

Habitually, tropical marine insects constantly are newcomer and have shortage of synchronized cohorts. In short-lived, summer-dry system, certain larvae of flies diapause in hyporheic sediments till surface movement proceeds.

Non-feeding and shortened adult life is characterized in Deuterophlebiidae, Nymphomyiidae and many Chironomidae. Deuterophlebiids have the shortest adult lifespan than any Diptera, with females living for a few hours and males possibly 30–45 min. A short-lived, and nonfeeding imaginal phase would be adaptive anywhere larvae get means for gamete making or where ecological conditions unfavorably affect adult persistence [52].

11. Life of Diptera

Most flies remain active throughout the year and many of them live less than a year. Many fly species survive the winter only as eggs. Others survive as pupae and a few survive as larvae or adults. Like all insects, they do not truly hibernate, but enter a state of diapause, which slows down their development and appetite, until temperatures rise and they become active again. Unless they hibernate, adult flies do not usually live very long, often only a month or two and sometimes just few days or weeks. Flies usually spend most of their lives as a larva or a pupa, and mostly spend the winter as adults in cracks and openings, and become active in spring. Flies are eaten by many predators, so very few of them live as long as they can.

12. Classification of Diptera

Diptera has worldwide distribution, diverse habitats and diets in both larvae as well as adults, while sizes range 1 mm–7.5 cm. Among differentiating taxonomic structures, wings are utmost distinct feature of Diptera, and these comprise a couple of functional front wings and condensed rear wings termed as halteres that help as balancing organs. With the exception of male scale insects, Diptera only have hind wings adapted into halteres. The thorax contains a complete mesothorax occupied with muscles that operate to forewings. The single couple of wings as well differentiate to Diptera from other insects so-called flies (dragonflies, caddisflies), whereas the posterior halteres isolate Diptera from other insects having single pair of wings (certain beetles and mayflies).

Separation of Diptera into suborders is established on wing venation and structure of antennae. Additional key features are chaetotaxy and arrangement of strong bristles in several fixed locations, and given specific or group names. Split-up of Diptera into families is based on habits (feeding), and habitats of adults and larvae. Species and genera are differentiated by details of head structure and profile of head, degree of separation and shape of eyes, and legs shape and proportions of segments. Abdominal shape generally defines distinctive appearance of a genus, however it is hard to express because the shape differs as the insect is starving, well fed or gravid (viviparous flies, for instance, tsetse fly) [53].

Diptera order is traditionally divided into two suborders distinguished by the differences in antennae, Nematocera (flies with multi-segmented antennae) and Brachycera (flies with stylate antennae) having about 110 families divided between them although one suborder Cyclorrhapha is non-monophyletic (flies with aristate antennae) [54].

The Nematocera species are recognized by their elongated bodies and many-segmented, often feathery antennae as represented by mosquitoes and crane flies.

The Brachycera have rounder bodies and much shorter antennae. The Nematocera comprises commonly delicate and small insects having lengthy antennae such as crane-flies, mosquitoes, midges and their relatives. The Brachycera contains more robust and compact flies with small antennae [55].

In older classifications of Brachycera, two Divisions have been recognized; Cyclorrhapha and Orthorrhapha. Orthorrhapha contains brachyceran flies devising obtect simple pupae, for instance, robber flies and horse flies, and Cyclorrhapha comprises brachyceran flies having enclosed pupae within tough puparium. Cyclorrhapha is additionally separated in two sets built on absence or presence of ptilinum and fissure linked to head. Ptilinum is an eversible pouch above antennae base used during emergence of adult fly to push on and open anterior end of puparium. The Aschiza has an absence of ptilinum and it though exists in Schizophora.

Nematocera in general are soft-bodied and slender flies having long antennae containing of several segments alike, palpi of many segments noticeably often drooping, and wings bear numerous longitudinal veins, however in the middle of wing generally lacking the conspicuous discal cell. When present, the anal cell is broadly exposed. Insects in Brachycera are generally fairly big flies, of stout body, antennae short, however occasionally showing traces of more than three segments, wings generally with a very thorough venation and with a discal cell, and palpi neither more than two-segmented nor conspicuously drooping. Cyclorrhapha includes the most highly specialized Diptera, mostly of short and stout build, with short antennae and many bristles. In recent decades by a suite of workers, the customary assemblages of Diptera have been analytically revised within a cladistic framework starting with the great dipterist Willi Hennig. Consent has emerged that several of traditional categories such as Orthorrhapha, Aschiza and Nematocera are not natural sets (they are paraphyletic). In other arguments these categories contain a group of basal lineages from that of other (monophyletic) categories (Brachycera, Cyclorrhapha and Schizophora) stand up. Recently, efforts to frame a monophyletic classification of Diptera have achieved pace, however to date, no overarching consensus has been gotten [56–58].

Order Diptera all together is a group of familiar insects that has traditionally been divided into three suborders [59–62]:-

12.1 Suborder Nematocera

Antennae contain flagellum, pedicel and scape having many segments alike; maxillary palpi bear in excess of three segments, frequently pendulous; anal cell open in wing; larvae generally with distinct head; mandibles opposed horizontally.

Family Tipulidae (crane flies or daddy long legs): Body, legs, wings elongated; slow-flying; larvae within soil (leather jackets), rotten wood, mud, moss, marine, fresh water, seaside.

Family Mycetophilidae (fungus gnats): Delicate, slim; dip in wet shaded dwellings, between rotting foliage.

Family Sciaridae (dark-winged fungus gnats): Related to fungus gnats, however extra dense, more frequently indoors.

Family Bibionidae (march flies): Solid, well-armored flies; spurs on legs strong; frequently plentiful over spring blossoms; larvae found within soil, on occasion tangled in a mass close to plant roots.

Family Scatopsidae (minute black scavenger flies): Analogous to march flies, more frequently indoors.

Family Cecidomyiidae (gall midges): Minute flies rarely seen as adults; larvae shapeless, tunnel in plant tissues, form plant galls, collapse foliage, stalks, roots; certain horticulture and agriculture pests.

Family Psychodidae (moth flies): Minute, wings hairy; frequently visible alone in kitchens, windows over sinks; larvae generally aquatic; several larvae plentiful in dirt sedimentation containers.

Family Phlebotomidae (sand flies): Thoroughly interrelated to Psychodidae; female adults blood sucker, transmit intestinal and dermal leishmaniasis, sand fly fever.

Family Ceratopogonidae (biting midges): Minute, usually wings spotted (*Culicoides*); female adults annoying by bite, blood sucker, transmit certain parasitic worms; *Forcipomyia* blood sucker of insects.

Family Chironomidae (nonbiting midges): Correlated to biting midges, however females not blood sucker; larvae marine; essential for fish food; adults fly close to water.

Family Simuliidae (black flies, buffalo gnats): Minute, humpbacked, antennae short; females blood sucker, transfer parasitic worms causing 'river blindness'; under skin form nodules; aquatic larvae, fastened to stones, freshwater crustaceans or underwater vegetation, filter feeders.

Family Culicidae (mosquitoes): Elongated; small; prominent proboscis; long palpi; recognized best through scaly wings; various females blood sucker, transport human diseases (Culicini transmit filariasis, yellow fever, viral encephalitis, dengue; Anophelini transfer malaria); aquatic larvae and pupae.

12.2 Suborder Brachycera

Antennae flagellum always mostly joined to a compound third segment, left over diminutive segments practice a bristle-like or stumpy style arista; wing anal cell narrowed, nearly usually closed on or earlier to wing border; palpi rarely have more than three segments, usually one or two, detained frontward (porrect); larvae head usually well-defined, mandibles travel parallel or vertically, may not be opposed; through a rectangular slit adults escape from pupa (Orthorrhapha).

Family Stratiomyidae (soldier flies): Colorful flies, found relaxing with wings closed on vegetation; males occasionally in air dance; sometimes larvae extended, active, carnivorous (*Stratiomys*), aquatic; others in decomposing foliage (*Hermetia*).

Family Rhagionidae (snipe flies): Unremarkable, relax on vegetation usually; certain females (*Symphoromyia*) blood sucker; maximum larvae in soil otherwise in water (*Atherix* females make egg-laying flights); certain form pits in soil, such as ant lions (*Vermileo*).

Family Pantophthalmidae (timber flies): Big, outmoded flies, found currently only in South American tropical forests; wood-boring larval grubs from time to time injury profitable wood.

Family Tabanidae (deer flies, horse flies): Short flies having large heads, eyes colored brilliantly; certain females (*Tabanus*, *Chrysops*, *Haematopota*) blood sucker, pests of livestock; several primitive genera only feed on flowers; larvae in wet soil or mud, either carnivorous (*Haematopota*, *Tabanus*) or vegetarian (*Chrysops*).

Family Asilidae (robber flies): Adults in flight clip other insects, suck blood; vary in sizes as of a few mm to 8 cm (lengthiest than other flies); distinctive 'moustache' of hairs possibly defends eyes from fly's victim damage; larvae in wood or soil; feed on several diet.

Family Bombyliidae (bee flies): Scaly, hairy; resemble to bees superficially, in similar way hover over flowers; often patterned brightly, by rubbing scales pattern ruined; in wasp and bee nests larvae scavenger otherwise parasitic (tsetse pupae, locust egg pods).

Family Scenopinidae (window flies): Black tiny flies, fond on indoor windows; larvae develop in carpets, nourish on clothes moth and flea larvae; usual habitation, nests of birds or related dry debris.

Family Therevidae (stiletto flies): Adults look like Asilidae, however not predacious; larvae elongated like Scenopinidae, worm-like, carnivorous however from time to time damage to plant roots.

Family Nemestrinidae (Tangle-veined flies): Somewhat similar to Bombyliidae; larvae parasitic on locusts, grasshoppers, beetles; noteworthy for attractive hovering pattern.

Family Acroceridae (balloon flies): Fantastic; tiny head; small thorax; swollen abdomen; larvae parasitic on spiders.

Family Empididae (dance flies): Adults take in insects blood, as well forage on flowers; *Hilara* projectiles above water, clasps minuet insects; larvae live in several habitations (decomposing vegetation, freshwater and marine dirt, flowing fluid from plants, fungi).

Family Dolichopodidae (long-legged flies): Minute, bristly, metallic flies; huge quantities assemble on foliage in misty locations; predacious on other insects; larvae similar to Empididae, lengthened, with slight outward head arrangement, identical habitations.

12.3 Suborder Brachycera-Cyclorrhapha

Generally condensed to Cyclorrhapha; typically make pupa within latter larval casing by way of a puparium; fly adults push off a rounded lid, therefore title as Cyclorrhapha; maximum families (Schizophora) have ptilinum (membranous pouch within head) that arises from horseshoe-shaped ptilinal suture (classifies Schizophora adults) over antennae, is puffed out and in to aid fly for escaping from puparium otherwise dust or to swell body of fly; ptilinum wastes and ptilinal suture leftovers only; individuals of minor group Aschiza, without ptilinal suture, are known primarily by their wing venation.

Series Aschiza: Do not possess a ptilinum, lack the prominent ptilinal suture on the face and have a puparium with a circular emergence opening, but not precisely ellipsoid.

Family Lonchopteridae (pointed-winged flies): Less well-known; famous for parthenogenesis; limited species globally; occasionally plentiful.

Family Phoridae (coffin flies): Minute flies, occasionally frequent indoors; larvae living in several biological fragments attractive with protein otherwise nitrogenous decaying produces; forage in nests of bees, wasps, termites, ants; breeding in carrion; several adults wingless or with small wings (brachypterous).

Family Pipunculidae (big-headed flies): Minute flies; head sphere-shaped; prominent for clear-cut hovering; larvae parasitic on Homoptera.

Family Platypezidae (flat-footed flies): Slight flies; legs peculiar; seen rarely; in wood fires smoke perform to dance; larvae living within fungi.

Family Syrphidae (hover flies): In wing vena spuria goes among third and fourth veins; aware in all places; fly above flowers, relax over foliage; certain larvae marine (rat-tailed' maggots); several species larvae forage on aphids living on plant leaves; stems.

Family Conopidae (thick-headed flies): Wasp like flies; larvae parasitic on wasps and bees; may possibly be an isolated evolutionary line.

Series Schizophora: Entirely flies in head have ptilinal suture, larvae without exterior head structure, through cuticle mouth hooks evident, one pair of prothoracic spiracles; one pair of posterior spiracles both have either a mass of small pores or three slits, larvae called maggots with hind end truncate; fore end pointed, both ends fleshy and blunt, with bulges; tracts of spines termed grubs.

Section Acalyptrata: Small soft-bodied flies; thoracic squamae (calypters which link base of wing to thorax) evanescent or small; key families established well; location of genera unclear; families may be gathered in relation to larvae diet likings.

Flies breeding in vegetable compost and dung:

Family Lauxaniidae (lauxaniid flies): Possibly, look to mimic other insects, feasibly leafhoppers; larvae live in decomposing plant material.

Family Helomyzidae (helomyzid flies): Alike to Lauxaniidae; larvae nourish on rotting animal and plant materials; most widespread of Acalyptrata.

Family Dryomyzidae (dryomyzid flies): Identical to Lauxaniidae, however have broader variety of diet, comprising fungi; yellow flies generally visible in winter.

Family Chyromyiidae (yellow flies): Length 1 or 2 mm; breed in mammal burrows, bird nests debris, cellars, caves; singly visible on indoor windows.

Family Celyphidae (beetle flies): Scutellum extremely expanded till it conceals wings and abdomen while at relaxation; dung breeding.

Family Mormotomyiidae (terrible hairy fly): Comprises single African species, wingless; appearance similar to spider; identified only from single area in Kenya; breeds in bat excrement.

Family Coelopidae (kelp flies, seaweed flies): Breed in wrack (decaying piles of seaweed held on seashores) primarily in moderate states; various species adults appealed by trichloroethylene; pests occasionally.

Flies breeding in animal refuse, dung, carrion:

Family Sepsidae (black scavenger flies, ensign flies): Minute, roundhead, black flies; occasionally spots at wing tips; can breed to invasion level in sewage sludge.

Family Piophilidae (cheese skippers): Larvae live in ham, cheese, dried fruits, cured meats, conserved pelts and skins; natural habitation is preserving carrion; named ‘skippers’ for the reason that larvae travel together by skipping and crawling (gripping tip of abdomen with mouth hooks and flipping body relatively through a long distance).

Family Micropezidae (stilt-legged flies): Big, tall-legged flies; patterned conspicuously blue-black wings; enormous in tropics.

Family Sphaeroceridae (small dung flies): Minute, black-brown flies; hind legs first tarsal segments inflated; plentiful all over world in dung like resources; on beaches certain members exist in seaweed; several wingless or short-winged species.

Family Sciomyzidae (marsh flies, snail-killing flies): Larvae aquatic, feed upon together dead and living snails; can be valued as controlling means for harmful snails.

Family Milichiidae (freeloader flies): Dung breeder; adults fasten to spiders and predacious insects and nourish on these; named ‘insect jackals’; *Madiza glabra* from time to time plentiful indoors.

Family Carnidae (bird flies, filth flies): Scavenge within burrows and nests; *Carnus hemapterus* adults scavenge between feathers of birds, breakdown wings.

Family Neottiophilidae (nest flies): Nest-breeding; larvae suck blood of nestling birds; resemble to calliphorid larvae.

Family Thyreophoridae (cheese flies): Among the rarest of flies; larvae are in dead bodies of large animals.

Family Chamaemyiidae (silver flies, aphid flies): Larvae predatory; recognized as controlling means of aphids and other soft-bodied insects including mealybugs and scales.

Family Braulidae (bee louse): Wingless fly, *Braula caeca*, exists in beehives; larvae feed on pollen and wax; adults attach to bee, can beg nutritious saliva similar to other bee colony members.

Flies with plant-feeding larvae:

Family Ephydridae (shore flies): Temporary; wide-ranging larval habitations; not any matter inedible to larvae (carrion, sewage, excrement, urine, brine tar pools, hot springs, algae); carnivorous petroleum fly (*Psilopa petrolei*) exists in crude petroleum pools seepage predating on confined insects; several larvae nourish on aquatic and terrestrial plants.

Family Diopsidae (stalkeyed flies): Certain larvae live in rotting plant matter, some mine in living vegetation.

Family Chloropidae (frit flies): Peak vital plant feeders; comprises pests of cereal and other crops.

Family Opomyzidae (opomyzid flies): Small, slender, yellow, brown or black colored flies; larvae live in stems of grasses including cereals.

Family Geomyzidae (geomyzid flies): Slightly slender; grayish or yellow color; seeing like pomace flies.

Family Psilidae (rust flies): Yellow to reddish, brown or black in color; head spherical; face often slanted backward; antennal third segment clearly elongated; carrot fly *Psila rosae* agricultural pest.

Family Agromyzidae (leaf miners): Larvae nourish in leaves parenchymatous tissue, reduce epidermis transparent and create either blotch mines or serpentine; deface decorative plants and shrubs.

Flies with fruit-feeding larvae:

Family Trypetidae (large fruit flies): Produce galls in several flowers mainly Compositae; a lot of Trypetidae larvae nourish in living fruits and deteriorate these; global scattering; damage by a number of members [Mediterranean fruit fly *Ceratitis capitata* (Wiedemann) is quarantine pest of fruits worldwide].

Family Drosophilidae (small fruit flies): Larvae live in decaying and fermenting fruits or sweet matter; comprises *Drosophila melanogaster*, used in genetic studies.

A number of minor families have been made to put up genera closely related to both above families, wherein, Otitidae (Ortalidae) and Lonchaeidae are most noticeably distinct, while others such as Pallopteridae, Ulidiidae, Camillidae, Diastatidae and Phytalmidae are unresolved.

Section Calyptrata: Characterized by large squamae (calypters which link base of wing to thorax); Scatophagidae are transitional.

Family Scatophagidae (dung flies): Living nearby dung, other rotting things; several as well predacious as adults and larvae.

Family Muscidae (housefly and allies): Several species including housefly; certain larvae specially in third instar carnivorous; breed in dung or decomposing plant material; *Fannia* larvae 'lesser housefly' like matters dipped in urine; economically significant muscid larvae nourish on plant roots and stems; subfamily Anthomyiidae comprises dipteran plant pests; stable fly *Stomoxys* (both sexes have biting proboscis) can be positioned in an isolated family Stomoxyidae; tsetse fly *Glossina* restricted to Africa, occasionally located in the family Glossinidae, occurred in North America.

Family Calliphoridae (blow flies): Certain bristly flies having carrion-feeding maggots; blow flies *Calliphora* (bluebottles) larvae feed in dead meat; *Lucilia* (greenbottles) occasionally invade living flesh; *Cochliomyia*, *Callitroga* (screw-worms) are hazardous feeders in living tissue.

Family Cuterebrinae (robust bot flies): Side-shoot of above family Calliphoridae; larvae parasitic in rodents; one larva *Dermatobia hominis* (human bot fly) as well attacks man; eggs on occasion attached to mosquitoes and other biting flies and passed to their potential victim.

Family Oestridae (bots and warbles): Larvae living in nose, under skin and in other head openings of big mammals; contains cattle warble fly *Hypoderma bovis,* sheep nostril fly *Oestrus ovis* and other species.

Family Gasterophilinae (horse bots): Larvae living in stomachs of horses, rhinos, zebras and elephants, involved to intestinal lining; association with other bot flies uncertain; presently categorized with other bot flies.

Family Sarcophagidae (flesh flies): Big, black and gray flies; common nearby garbage holes; larval behaviors varied, found in dead or living animal material; several viviparous species.

Family Tachinidae (tachinid flies): Biologically essential in equilibrium of nature for the reason that larvae are parasites in woodlice, other insects, centipedes, spiders; employed in biological pests control.

Section Pupipara: Doubtful group, families can only be convergent in habit; lay living larvae; adults of both sexes exclusively feed on blood.

Family Hippoboscidae (louse flies): adults feed on blood of birds and mammals; several fly, certain have wings lost or reduced (sheep ked *Melophagus ovinus*).

Family Streblidae (bat flies): Distinctive rounded head; wings generally efficient however fly slightly; closely cling to host.

Family Nycteribiidae (wingless bat flies): Continuously wingless; weakened and de-sclerotized thorax; living completely on bats; hardly detectable as flies.

13. Damage caused by Diptera to cured fish

Some Diptera cause significant damage in many ways, generally during the larval stages. The feeding by larvae of Calliphoridae causes quantitative losses on moist fish. These injuries can be severe if circumstances are optimum for fly growth and under such conditions, i.e., if poorly or unsalted and salted fish are dried gradually for the reason that of rain or high humidity, weight harms of 10–30% by fly larvae may be caused. Disintegration of fish by fly invasion can cause quality damage and may lead to bigger danger of damage by mites and beetles. Significant weight losses because of fragmentation of fish during treating have been noted, however the involvement of blow fly injury to this has not been assessed separately. Thorough and heavy salting provides complete protection against blow fly larvae.

Most flies found on cured fish belong to the subfamilies Calliphoridae (blowflies, bluebottles, greenbottles, screw-worms including *Calliphora* sp., *Chrysomya albiceps* Wiedemann, *Chrysomya bezziana* Villeneuve, *C. chloropyga putoria* (Wiedemann), *C. megacephala* (Fabr.), *C. regalis* Desvoidy, *Lucilia cuprina* Wiedemann, and *L. sericata*) and Sarcophagidae. Flesh flies includes *Sarcophaga* sp., *S. nodosa* Engel, *S. tibialis* Macquart and *Wohlfahrtia* sp.), in Calliphoridae family. The most common of these are several species of Chrysomya; however, Lucilia, Calliphora, Wohlfahrtia and Sarcophaga have too been stated. Other families of flies infesting cured fish denoted in records are Phoridae (Megaselia), Muscidae (Musca, Atherigona and Ophyra), Piophilidae (Piophila), Ephydridae (Discomyza) and Milichidae (Leptometopa). Entirely, these flies are somewhat alike in general form, although they display a diversity of coloration and size, and their proof of identity needs specialist information [63].

The larvae of some species can cause myiasis in livestock or man i.e., they may infect external wounds or can be swallowed and carry on developing in the intestine as parasites. Adult flies of many pest species are invited to decomposing material (such as rotting offal of fish) and dung, wherever they may breed and feed. They might thus spread pathogenic organisms when these lay eggs on fish.

14. Biting Diptera

Biting flies are two-winged external insects that feed actively on the blood of vertebrate hosts in the morning or evening and at night or day, and their biting is of a considerable nuisance. Their irritating bites could transmit pathogenic organisms that cause devastating loss of human and animal lives. The biting insects suck blood from humans and animals, and their biting is of a significant annoyance. More importantly, they are carriers for a number of organisms producing diseases and result in expiries on a huge scale. The most significant set of biting Diptera is mosquitos that have a slender and long body, and needle-shaped long piercing mouthparts. Others comprise phlebotomine sandflies, blackflies, tsetse flies, biting midges, stable flies and horseflies (tabanids), which normally have smaller biting mouthparts and additional robust bodies. The last three sets as vectors of human disease are of limited importance [64].

14.1 Mosquitoes (Culicidae)

Mosquitos diverge from other biting Diptera in having long needle-shaped mouthparts, a long slender body and long legs. The wings occasionally have noticeable outlines of scales. The adult insects measure between 2 and 12.5 mm in length. Certain species bite at night or in morning and evening, whereas others feed out of doors or during the day time indoors. There are several important genera of mosquitoes and key genera include *Culex, Aedes* and *Anopheles.*

Males of the numerous species do not suck blood but feed on plant juices. The females usually mate only once, but produce eggs at intervals throughout their life and so most female mosquitos require a blood-meal. The ingestion of a blood-meal and the coinciding eggs development take 2–3 days in tropics, however longer in temperate regions. The gravid females look for appropriate places to lay their eggs, afterwards which another blood-meal is taken and another batch of eggs is laid. This practice is repetitive till the mosquito perishes. The mosquito life cycle involves eggs that are laid mainly in water. In some species, eggs are laid signally, while in others, these are laid joined together in rafts. Dependent on the species, a female lays eggs between 30 and 300 at a time. Various species directly lay their eggs on water surface either singly (*Anopheles*) otherwise fixed jointly in floating rafts (*Culex*). Particular species (*Aedes*) lay their eggs just above the water link otherwise on wet mud and only when flooded with water these eggs hatch. If left dry, these can keep on viable for several weeks. Larvae (wigglers) hatch and feed on aquatic material, pupate (become tumblers) and eventually emerge as adults. Adult females may live for several months [65, 66].

Among the mosquitos there are two groups that suck human blood and may transmit disease. The anophelines; the genus *Anopheles* is best known for its role in transmitting of malaria, but in some areas it can also transmit filariasis. The culicines comprise the genera *Aedes,* vectors of dengue, yellow fever and other viral diseases and from time to time of filariasis; *Culex,* vectors of filariasis and various viral diseases; *Mansonia*, transmitter of *brugian filariasis*; and *Sabethes* and *Haemagogus*, spreaders of yellow fever in forests of Central and South America. Mosquitos *Aedes*, *Anopheles* and *Culex* may be differentiated from each one by way as presented in **Figures 8–12**. The best valuable features to distinguish anophelines from other

DOI: http://dx.doi.org/10.5772/intechopen.91391

mosquitos are size of palps that is equivalent to proboscis; whereas these generally at rest retain their abdomen and mouthparts in a straightforward stripe at an angle to relaxing place; the angle differs within species and in various circumstances it is nearly erect to surface. In south Asia, vector of malaria *Anopheles culicifacies* Giles, is an exemption by keeping body nearly equivalent to surface. As per its name proposes, this one superficially looks similar to mosquito *Culex*. A number of mosquitoes, including *Psorophora*, *Aedes*, *Mansonia*, *Culex* and *Anopheles* spp., are important parasites of domestic animals. When the blood-feeding females are present in large numbers, they cause stress to animals and produce severe anemia [67, 68].

14.1.1 Anopheles *mosquitos*

About 380 species of *Anopheles* mosquitoes occur around the world. Some 60 species are sufficiently attracted to humans to act as vectors of malaria. A number of *Anopheles* species are also vectors of lymphatic filariasis and viral diseases. Female and male adult mosquitoes may be distinguished from every one due to occurrence of antennae that is bushy in males than females. Genus *Anopheles* may be distinguish from whole others genera based on occurrence of clubbed palps within males, while non-clubbed in rest of species and lengthy palps in *Anopheles* female whereas small palps in all other females (**Figure** 7).

The most preferred breeding sites are pools, seepages, quiet places in slow-running streams, rice fields, leaf axils of certain epiphytic plants and puddles of rainwater, but not artificial containers, except in the case of *Anopheles stephensi* Liston (**Figure 8**). The eggs are elongated, about 1 mm in length, have a pair of

Figure 7.
(a) Male antenna, (b) Female antenna.

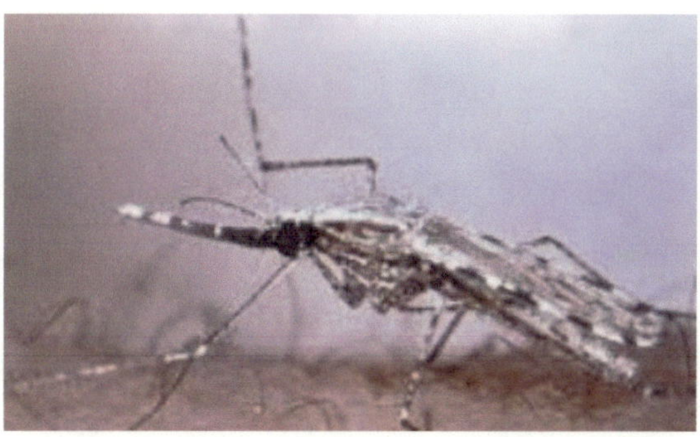

Figure 8.
Anopheles stephensi.

lateral floats and laid singly on the water surface where these float until hatching that occurs in 2–3 days. The larvae float in a horizontal position at the surface, where these feed on small organic particles. In the tropics, the duration of development from egg to adult is 11–13 days. Some species feed mostly on animals, while others feed almost entirely on humans. Mosquitos *Anopheles*, are active between sunset and sunrise, and there are variations also in their liking for biting outdoors or indoors. The anophelines, which come in houses to feed habitually rest for a few hours indoors after feeding, then can leave for outdoor protected resting sites like burrows, vegetations, crevices and cracks in ground, trees or in caves and undersides of bridges. On the other hand, they may stay for the whole period indoors required to digest blood-meal and produce eggs. Once the eggs are fully developed the gravid mosquitos leave their resting sites and try to find a suitable breeding habitat [69–72].

14.1.2 Culex mosquitos

About 550 species of *Culex* have been described, and most of them are from tropical and subtropical regions. Some species are important as vectors of bancroftian filariasis and arboviral diseases, such as Japanese encephalitis. Rafts of 100 or more eggs are laid on the water surfaces that remain afloat until hatching occurs 2–3 days later. These breed in a large variety of stagnant waters, ranging from artificial containers to large bodies of permanent water. The most common species, *Culex quinquefasciatus* Say (**Figure 9**), is a major nuisance and vector of bancroftian filariasis, which breeds especially in water polluted with organic material, such as refuse and excreta or rotting plants, pit latrines, blocked drains, canals and abandoned wells. It is markedly a home species, and females bite to people and animals all over the night outdoors and indoors. For the period of day, these are not active and frequently found hidden in dark corners of rooms, and as well rest outdoors in holes within trees and on vegetation areas. The vector of Japanese encephalitis, *Culex tritaeniorhynchus* Giles (**Figure 10**), prefers cleaner water and is most commonly found in irrigated rice fields and ditches [73].

14.1.3 Aedes mosquitos

Mosquitoes *Aedes* occur around the world and these can cause a severe biting annoyance to people and animals, both in cooler climates and in the tropics. In tropical countries, yellow fever mosquito, *Aedes aegypti* (**Figure 11**) is a key vector of yellow fever, dengue fever, dengue hemorrhagic fever and other viral diseases. A thoroughly related species, Asian tiger mosquito *Aedes albopictus* (Skuse),

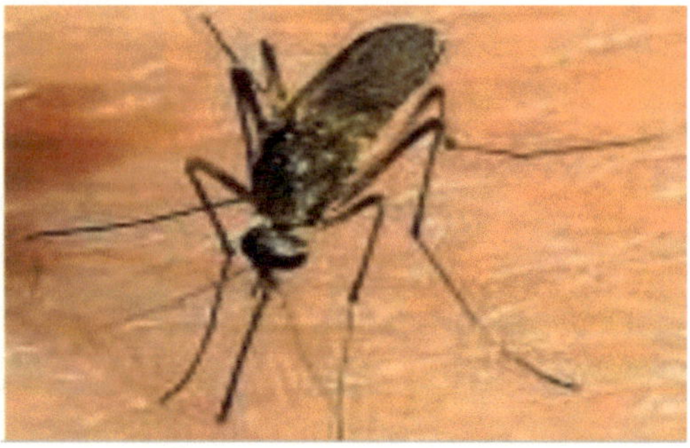

Figure 9.
Culex quinquefasciatus.

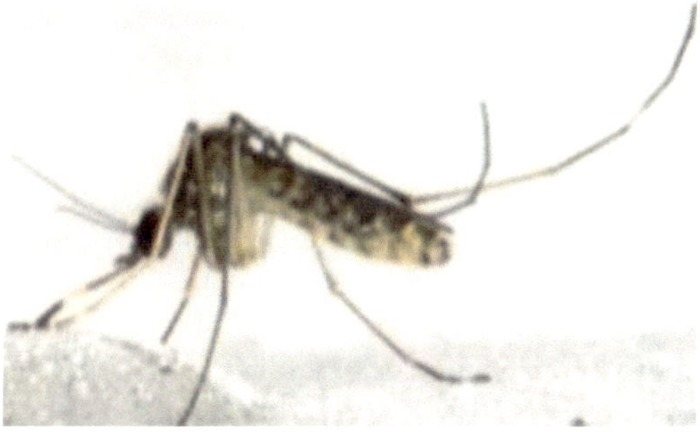

Figure 10.
Culex tritaeniorhynchus.

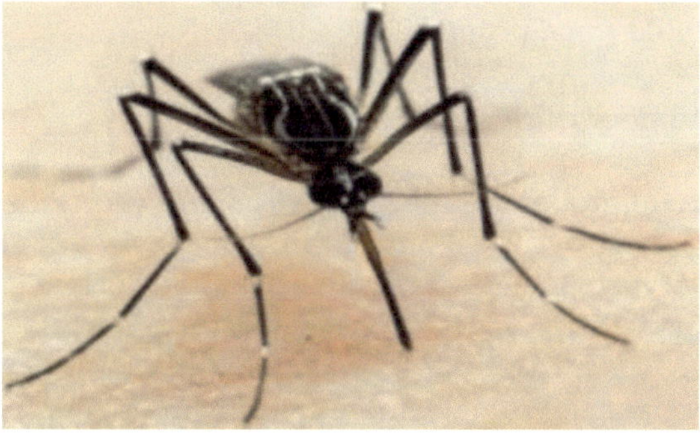

Figure 11.
Aedes aegypti.

can spread dengue as well (**Figure 12**). In certain regions *Aedes* species also spread filariasis [74–77].

14.1.4 Mansonia mosquitos

It includes 15 species classified in subgenera *Mansonia* Blanchard and mosquitos are mostly found in marshy areas in tropical countries. These mosquitoes can be very alike to common mosquitoes in genus *Aedes*, however vary by having the last segment of abdomen much broad, rather than markedly narrower as in *Aedes*, and through having most of scales very broad on the top of wings, rather than long and narrow as these are on *Aedes*. These may as well be differentiated from several other mosquitos genera through having a tuft of bristles (post-spiracular bristles) on body plate instantly

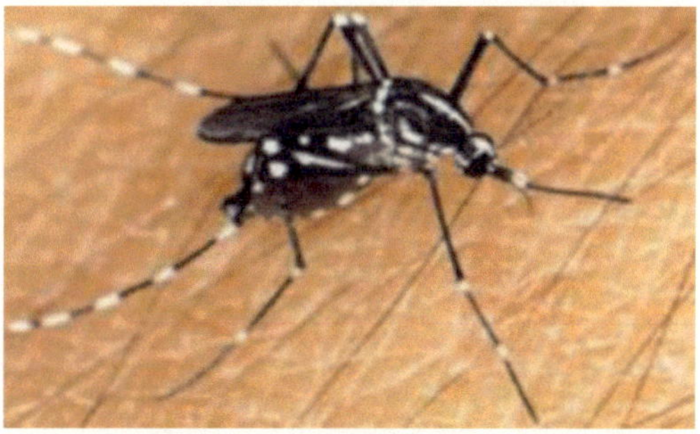

Figure 12.
Aedes albopictus.

behind spiracle on the sideways of thorax. Some species are important as vectors of the helminthes that cause brugian filariasis. Their body, including the legs and wings, is covered with dark-brown and pale scales, giving it a rather dusty appearance, as if sprinkled with salt and pepper. The larvae of *Mansonia* species occur in permanent waters in association with aquatic plants that have roots used for attachment by the siphon to obtain oxygen for respiration from air cells. Mosquito *Mansonia titillans* (Walker) (**Figure 13**) is known to transmit various arboviruses, including Venezuelan equine encephalitis, while species *Mansonia dyari* Belkin, Heinemann and Page should be considered a potential vector of Rift Valley fever virus [78].

A number of control approaches are employed against every stage in the life cycle of mosquito. Difficulties exist with the whole forms of control options and their continuous applications are generally required to produce any effect. Different forms of natural control are currently being trialed with personal protection of human, and selecting and breeding of livestock species, which are more resistant to mosquito biting to hold some promise.

14.2 Horse flies and deer flies (tabanids)

The most important groups are the genera *Tabanus* (horseflies, greenheads), *Chrysops* (deerflies, mangrove flies) and *Haematopota* (clegs or stouts). They are of minor importance as vectors of diseases, such as tularemia and certain arboviral diseases. In West and Central Africa, some species of the genus *Chrysops* transmit the filarial parasite *Loa loa*. Horse flies can transmit anthrax, anaplasmosis, some trypanosomes, tularemia and some helminthes infections. The tabanids are robust and strong fliers and they measure about 5–25 mm in length. They have a big head with apparent eyes that show iridescent colors. The mouthparts are big and pointed downward. The wings are clear entirely or have brownish color or spots. Wings are folded flat along the body when insect is at res. Females horse fly *Tabanus*

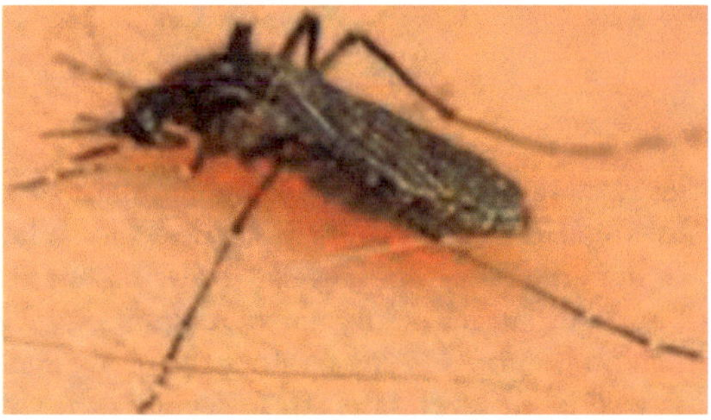

Figure 13.
Mansonia titillans.

Figure 14.
Tabanus trimaculatus.

trimaculatus Palisot de Beauvois (**Figure 14**) have scissor-like mouthparts that aim to cut the skin and then lap up the blood. Deer fly *Chrysops callidus* Osten Sacken (**Figure 15**), is blood sucking insects of humans and cattle. They are large flies with large brightly-colored compound eyes, and large clear wings with dark bands. They are in habit of hovering around the head and shoulders and biting people, and known for following moving animals and repeatedly buzzing the head [79].

This family (Tabanidae) comprises deer flies and horse flies. Still, the life cycle of several species is unfamiliar, however in those that are identified, there is often an aquatic phase in the cycle. Eggs are placed in moist areas including pools and streams. In about a week, larvae hatch out and stay in the bottom of pools habitually burrowing into mud where these nourish on various forms of biological material and often surviving as micro-predators. Several species hibernate in winter as larval stage, however in spring; larvae molt to a pupal stage (lasting for 2–3 weeks) and then adults emerge. The female tabanids are only blood suckers. Mouth parts in many species are fairly prominent and role like a spear otherwise stylet to puncture an area, and usually causing loss of blood afterward the fly has done nourishing. An adult female tabanid may take away nearly 0.2 cc of blood for each nourishing. Their bites are deep and painful, and the wounds often continue to bleed after the flies have left host [80].

14.3 Stable fly *Stomoxys calcitrans* (Linnaeus)

Stable flies (*Stomoxys*) occur around the world. Stable flies are dark, medium-sized flies, 5–6 mm in length, resembling house flies in shape and size, and are also

Figure 15.
Chrysops callidus.

Figure 16.
Stomoxys calcitrans.

known as biting house flies. They can be distinguished from house flies and other similar looking flies by their forwarded pointing mouthparts. They may be confused with tsetse flies (Glossina), and these can be distinguished from tsetse flies, which also have forward-pointing mouthparts, by their smaller size and the position of wings, which do not overlap at the back in stable flies when at rest. These create painful bites and are a severe annoyance to humans and animals. They are not significant as vectors of diseases. But, they occasionally play a part in spread of myiasis by carrying eggs of myiasis-producing fly *Dermatobia hominis* (Linnaeus) [81].

Fly *Stomoxys calcitrans* (**Figure 16**), is commonly called the stable fly, barn fly, biting house fly, dog fly or power mower fly. Unlike most members of the family Muscidae, *S. calcitrans* (sharp mouth plus kicking) and others of its genus suck blood from mammals. The eggs are put down in decomposing vegetation or manure and in the same location, larval and pupal stages progress by taking nearly 3 weeks to complete life cycle. Equally, females and males are blood feeders and severe pests of both animals and man. The larvae are creamy white in color and resemble those of the house fly. The pupae develop in dry areas in the soil. Development from egg to adult takes from 12 days to 2 months, depending on the temperature. They are not as important as vectors of disease; however, stable fly can transmit many organisms including anaplasmosis, anthrax and other blood-borne organisms [82].

14.4 Horn fly *Haematobia irritans* (Linnaeus)

Although a small fly, the adult of this species (Muscidae) is one of the most important ectoparasites of pastured cattle. Adults are half the size of a house fly (7 mm), gray in color with the large compound eyes and reduced antenna (**Figure 17**). In the life cycle, the eggs are laid on newly passed feces. Larvae are approximately 7 mm long, of pale yellow color and with a simple elongate body that lacks a sclerotized head. In that atmosphere, the complete life cycle takes place, generally taking about 2 weeks for completion. Typically, the adults exist in great numbers along the withers, base of the horns and caudal folds. The adults stay on animal for the whole time (excluding when eggs are being set down), however they feed only once or twice a day and males and females both are blood feeders. The adults fly typically takes position by face downward when sitting on an animal. Besides loss of blood along with heavy constant invasion, this fly looks to cause an excessive annoyance and irritation. Infection conduction comprises anaplasmosis and other blood-borne organisms.

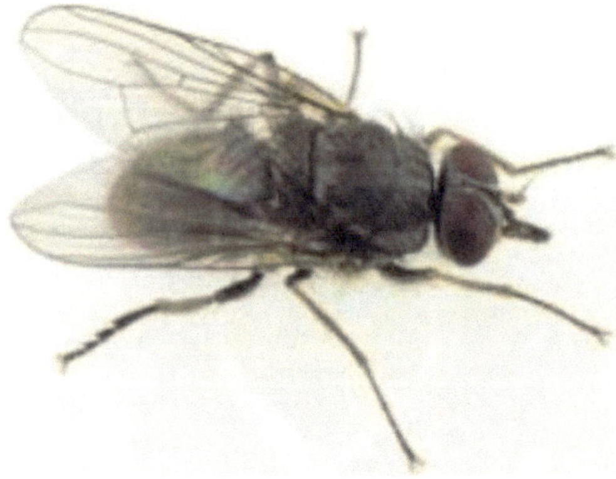

Figure 17.
Haematobia irritans.

Managing of horn fly is commonly factual as soon as compost is either dried or often removed to break life sequence. Further operational control actions include anti-larvicidal mixtures, which are added either to diet or delivered as supplement. In latter cases, action must be on track before start of fly season [83].

14.5 Tsetse flies *Glossina* sp.

Tsetse flies occur only in tropical Africa and include all the species in the genus *Glossina*, which are placed in their own family Glossinidae. They are yellowish or dark brown, medium-sized flies and 6–15 mm in length. They can be distinguished from other large biting Diptera by their forward-pointing mouthparts. Tsetse flies bend their wings completely once they are relaxing so that one wing directly rests on top of other above their abdomens. Tsetse fly has also a long proboscis that spreads directly frontward and is attached to bottom of its head by a distinct bulb. Both sexual category are blood feeders and nourish on a wide-ranging of hosts also comprising man. Eggs hatch in the body of female and later on larval growth takes place, and then dropped larvae pupate instantly. The pupal stage nearly lasts for 3 weeks. This fly is a powerful vector for some diseases of man and animals comprising trypanosomiasis. They have a prominent economic impact in sub-Saharan Africa as the biological vectors of trypanosomes, which cause human sleeping sickness and animal trypanosomiasis, for instance, *Glossina brevipalpis* (Newstead) (**Figure 18**) is a vector of *Trypanosoma congolense* and *Trypanosoma vivax* [84].

14.6 Black flies

A black fly sometimes called a buffalo gnat, turkey gnat or white socks, is any member of the family Simuliidae. Black flies occur around the world and there are about 1300 species in the genus *Simulium* (**Figure 19**). The buffalo gnat blackflies are stout-bodied, about 1–5 mm long, and usually black in color, although orange and yellow species exist as well. They have relatively large eyes, legs are short, and the wings are short, broad and colorless. Black fly bites in daytime out of doors and some species prefer to feed only on certain parts of the body, for example, the legs or the upper part of the body. Black flies prefer to lay eggs in swift oxygen-rich running water in streams, rivers and spillways of dams, and are deposited on objects or near the surface of the water. In the tropics, the eggs usually hatch after 14 days. After the eggs hatch (1–4 weeks), larvae attach to the submerged objects and remain there for about a month. The larvae do not swim (usually symptomless), continuing attached to submerged flora, stones and other substrates, and forage on tiny suspended particles. Based upon climate, the larval stage persists from 1 week to many months. The pupae are as well attached to submerged things

Figure 18.
Glossina brevipalpis.

Figure 19.
Simulium *black fly.*

and when the larval stage pupates, adults develop in about a week. Although small in size, the adults are good flyers and may go several kilometers in search of food [85].

Onchocerciasis, too known as river blindness, is a sickness initiated by infection by the parasitic nematode *Onchocerca volvulus*. Warning signs comprise intense itching, bumps below the skin and loss of sight. The parasite worm is transported by the bites of a *Simulium* type black fly. Only adult females are blood feeders, wherein the best significant species are *Simulium neavei* Roubaud and members of *Simulium damnosum* Theobald complex. Additionally, black flies are of severe annoyance in numerous regions of the world for the reason that of their painful bites and from time to time huge numbers involved in invasions. Black fly bites can create localized inflammation and swelling, and extreme irritation of skin lasting for days or weeks. Usually, black flies do not come into houses, but bite in the daytime and outdoors, especially along riverbanks. Certain species show a strong preference for biting specific parts of the body, for example, *S. damnosum* in West Africa mainly attacks the legs. Most species feed predominantly on birds or mammals, while several feed on humans [86].

14.7 Phlebotomine sandflies

Sand flies (Psychodidae) are about 1.5–4 mm long, have a hairy exterior, visible black eyes and lengthy stilt-like legs. They have a typical jumping flight with several short flights and landings. Contrary to all other biting Diptera, when at rest, their wings are held upright above the body. Sand flies are minute blood sucking flies that are key as vectors of leishmaniasis and can cause a severe biting nuisance, but limited to a small area. Species that take place in the Mediterranean area can spread sand fly fever that is a viral disease likewise recognized as pappataci fever or 3-day fever. The breeding places for this genus appear to be mainly non-aquatic situations. The life cycle may last from 1 to 4 months, depending on species and temperature, although it usually lasts less than 45 days. Sand flies feed on plant juices, but mostly the females need a blood meal in order to develop eggs. The cattle provide an abundant source of blood, while the stables and houses provide suitable resting places. Blood is taken from humans and animals such as dogs, farm livestock, wild rodents, snakes, lizards and birds [87].

The sand fly *Phlebotomus papatasi* (Scopoli) (**Figure 20**), is the main vector of the Old World cutaneous leishmaniasis. It is distributed from Morocco to the Indian subcontinent and from southern Europe to central and eastern Africa [88].

14.8 Biting midges

Biting midges (no-seeums, punkies) are blood sucking flies and about 1.5 mm in length. The most important genus *Culicoides*, is distributed worldwide and can cause a serious biting problem, as can the genus *Leptoconops*. These insects are vectors of the human filariae parasites *Mansonella ozzardi* and *Mansonella perstans* (Nematoda: Onchocercidae) that are mostly deliberated to be not hurtful to humans. These insects are termed as sand flies in certain regions of the world, however these may be differentiated from phlebotomine sand flies using the point that while midges are at relaxation wings are bent uniform on body; moreover, these frequently fly in crowds nearby head or other uncovered parts of body and do not flutter in a hopping mode by several landings and short flights, as is done by phlebotomine sand flies.

The lifecycle of this genus involves aquatic breeding places, elevated surface of mud or wet soil primarily temporary pools, decaying leaf litter and objects near or partially in water. The larvae feed on decaying organic matter and the time taken for development from egg to adult may be 2–4 weeks. It is an important vector for blue-tongue virus in many animal species as well as being a pest. Individual midges can cause a painful bite, but they are considered to be an especially severe pest because of their habit of attacking in swarms of hundreds or thousands. Ceratopogonidae is an example of family that includes serious blood-sucking *Culicoides sonorensis* Wirth & Jones (**Figure 21**) feeding both on humans and

Figure 20.
Phlebotomus papatasi.

Figure 21.
Culicoides sonorensis.

Figure 22.
Melophagus ovinus.

other mammals, and spread the livestock diseases blue tongue and African horse sickness [89].

14.9 Sheep Ked or sheep tick *Melophagus ovinus* Linnaeus

It is a fly from the family Hippoboscidae, brown and hairy in color and resembles a tick. This wingless fly is about 4–6 mm long and has a small head. They are blood-feeding parasites of sheep. The sheep ked feeds on the blood of host by inserting its sharp mouthparts into capillaries beneath the skin. The adult hippoboscids are well adapted to an existence on wool, hair and feathers for blood feeding. The life span of this fly is about 4 months with adult females retaining larvae internally until pupation and may produce 10–20 larvae by producing a single larva at a time. Pupae are attached directly to the wool, pupal stage lasts for 19–23 days and adult lives for 7–10 days. The entire life cycle of this fly takes place while it is on the infested animal. Other hippoboscid flies are important vectors for some avian diseases such as *Haemoproteus* sp. The sheep ked *Melophagus ovinus* (**Figure 22**), once remained a serious pest in the sheep industry. Several similar genera are present on wildlife including deer and elk. Adults ked can be killed using treatment dips and sprays most commonly containing ivermectin or pyrethrin. Use of injectable antihelminthics is also effective against some arthropods that have made this a less frequent problem [90].

15. Myiasis-producing Diptera

Myiasis is the invasion of tissue by fly larvae, which at least for a certain period, feed on the host's dead or living tissue, liquid body-substance or ingested food and such invasions can be benign in effect, but others may result in a variety of conditions, including death. When the invasion occurs in the intestinal tract, it is called intestinal myiasis, in stomach known as gastric myiasis, or there may be nasal myiasis and cutaneous myiasis, etc. Some species of flies that are not significant as adults are important as myiasis-producing larvae. Cutaneous myiasis is a skin invasion by larvae (maggots) of certain flies, and depending on the species of fly involved, there are three main types of skin infestation by fly larvae such as furuncular (pimple-like or boil-like) myiasis, wound myiasis and migratory myiasis [91].

Many of the flies that cause furuncular myiasis are commonly known as bot flies including *Dermatobia hominis*, *Cordylobia anthropophaga* (Blanchard & Berenger-Feraud), *Wohlfahrtia vigil* (Walker) and *Cuterebra* species. Many of the flies do not lay eggs on humans. Instead, the flies lay their eggs on other insects (such as mosquitoes) or on objects (such as drying laundry) that may come into contact with people's skin. Characteristic signs of furuncular myiasis comprise irritation, an impression of movement and on occasion a sharp piercing aching. In the beginning, people have a minor red bump, which can look like a common insect bite or creation of a furuncle (pimple). Later on, bump expands, and a slight opening may be evident in the center, opening may drain a clear yellowish fluid and on occasion a small portion of larva end is evident. Open wounds (traumatic), typically in homeless people, alcoholics and others in poor social circumstances, may become infested with fly larvae. The tissues that line the mouth, nose or eyes (mucosa) may also become infested [92]. The most common flies are screwworm flies such as *Cochliomyia hominivorax* and *Chrysomya bezziana* and *Wohlfahrtia magnifica* (Schiner) (**Figure 23**).

The most common sources of migratory (creeping) myiasis are flies that typically infest horses and cattle (Gasterophilus and Hypoderma flies). People can become infested if they have contact with infected animals. Less often, the flies lay eggs directly on people. Larvae do not stay in one spot and they burrow under the skin, causing itchy lesions. Diagnosis of fly larvae can be made on the size of larvae, location and host from which it is recovered, and characteristics of the spiracle openings located on the posterior of the larvae [93].

Warble flies of cattle are perhaps one of the most significant myiasis-producing problems of the cattle industry. There are two important species of the cattle grub, *Hypoderma lineatum* De Villiers Southern cattle grub and *Hypoderma bovis* (Linnaeus) Northern cattle grub. In broad-spectrum, life sequence of species contains adult flies that deposit eggs on body of host by sticking to furs, within some days eggs hatch and larvae enter into skin. Afterward, larvae transfer through host tissue and gather at esophagus wall. These from here, move to dermal tissue of host back, persist on cattle back for quite a few months and after that time these drop out to land for pupation. There are a number of indirect and direct reasons of hurt and fiscal loss which may be ascribed to invasions by these larvae. Mortalities may take place for the period of larval movement and at what time these are assembled in an area of esophagus. Indirect injuries are described as condensed milk making, weight loss otherwise low weight increases and less price of hide or carcass owing to presence of larvae. Common warble of rodents and rabbits *Cuterebra* sp., although common in wild rodents, infestations with this larvae are also occasionally seen on

Figure 23.
Wohlfahrtia magnifica.

dogs and cats. One species (*Cuterebra emasculator*) Fitch parasitizes the external reproductive organs of rodents and may have an effect on population numbers [94].

Sheep bots or head grub *Oestrus ovis* L., (**Figure 24**) of family Oestridae, is a bot which commonly enters the sinuses nasal and passages of sheep. Adults of fly place live larvae over nasal passages; from here these make their approach to frontal sinuses and attach themselves with mucus membranes. These persist for some months in this location and are sneezed out or ultimately fall out and pupate in soil. The pupae period lasts for around 3–6 weeks. When host is infested; generally there is a pus-filled ejection from the nostrils, resulting shaking of head vigorously, less appetite and grating of teeth by animal. Maximum of cases do not produce distinct clinical symbols even though expiry of animal may take place within a week after intensified signs. The indirect damages and serious situations causing from invasion are possibly the best reasons for starting treatment [95].

Horse stomach bot *Gasterophilus* sp., in the family Oestridae are several species of this parasite each of which has slightly different locations of attachment, primarily affecting horses and donkeys. The more common horse bot fly *Gasterophilus intestinalis* (DeGeer) (**Figure 25**) is an internal parasite of gastrointestinal tract, while other are *Gasterophilus nasalis* (Linnaeus) nose bot fly and *Gasterophilus haemorrhoidalis* (Linnaeus) throat bot fly. In mid-summer, adult flies are frequent and females lay eggs on hairs habitually on the belly, forelegs, inside the knees and flanks. When the horse licks these body parts, rubbing and moisture cause eggs to hatch. Larvae transfer to mucus membrane of lips and tongue ultimately creating their way to stomach or other places. At this point, these continue to attach till subsequent spring, as soon as these separate, pass out through feces, fall to ground and pupate. Some indirect and direct pathological modifications are linked to invasion. Modest invasions by bots can contribute with no apparent symbols, but strong invasions may be shown through intestinal syndromes. Probable interfering with ingestion and immersion of nutrients, irritation of mucus membranes, and obstacle of pyloric sphincter are very common [96].

Figure 24.
Sheep bot fly. Immature and mature larvae.

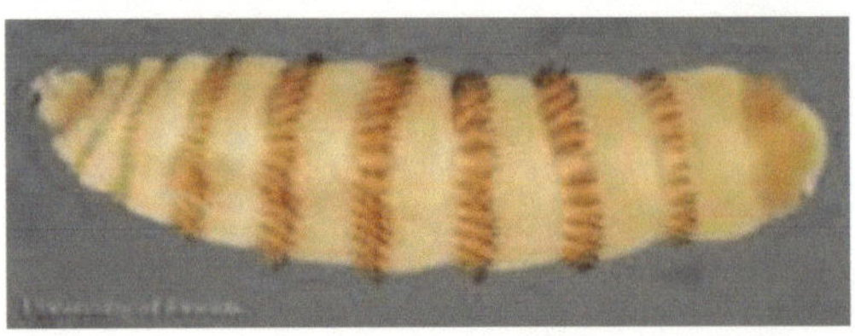

Figure 25.
Gasterophilus intestinalis *larva.*

15.1 Cutaneous myiasis

Genus *Cochliomyia* in the family Calliphoridae includes blow fly *Cochliomyia hominivorax* (Coquerel) well-known as primary screwworm (**Figure 26**) for the reason that this larva creates myiasis and forages on live tissue producing pocket-like deep lesions in skin that might be enough harmful to host animals. But, *Cochliomyia macellaria* (Fabricius) is famous as secondary screwworm since its larva creates myiasis, however nourish on only necrotic tissue. Forensically, that species is significant since it is habitually linked o dead bodies and carcasses. All species in the family Calliphoridae have bristles on their merones, plumose arista and well-developed calypters. Both *C. macellaria* and *C. hominivorax* are metallic green to bluish green in major coloration and three black longitudinal stripes (vittae) on the notum of the thorax. The species *C. macellaria* has pale setulae on the fronto-orbital plate outside the row of frontal bristles, while *C. hominivorax* has dark setulae. The larvae of both *C. macellaria* and *C. hominivorax* have cylindrical bodies tapering anteriorly with 10 or more robust spines around the spiracular area and bands of small spines on each segment. The *C. hominivorax* larvae have distinctly pigmented tracheal trunks, while *C. macellaria* larvae do not have pigmented tracheal trunks, but bear spines in a V shape on the anal protuberance [97].

The gravid female screw worm fly is captivated on living animals to oviposition sites. These sites are any discharges, bites, wounds, etc., which may take place. For egg deposition, the naval of newborn animals is a common site. The eggs are of cream color, hatch in 24 hours and larvae enter the wound and begin feeding. The larvae burrow into tissue, enlarging the wound that cause severe pain to the host. Animals smaller than rabbits, usually do not survive due to infestations. Larger animals may surrender to repeated infestation or if larvae penetrate blood vessels. Death is usually caused by toxemia and or septicemia from bacterial invasion of the wound. After 5–7 days, the larvae drop, burrow into the first layer of topsoil and begin their pupation. This stage can last from 7 days at a warm temperature to as long as 2 months if the weather is much colder. After emerging from the pupa adult flies live around 2–3 weeks. Once the infestation commences, a dark brown or reddish-brown discharge begins leaking from the wound, sometimes accompanied by an unpleasant smell as the flesh begins to decay both in livestock and human victims [98].

These are various important species of parasites and necrotic tissue feeders since they are common and capable to parasitize abrasions and wounds on animals. Blow fly strike is multiparty, such as on sheep and some other animals. Altogether, these

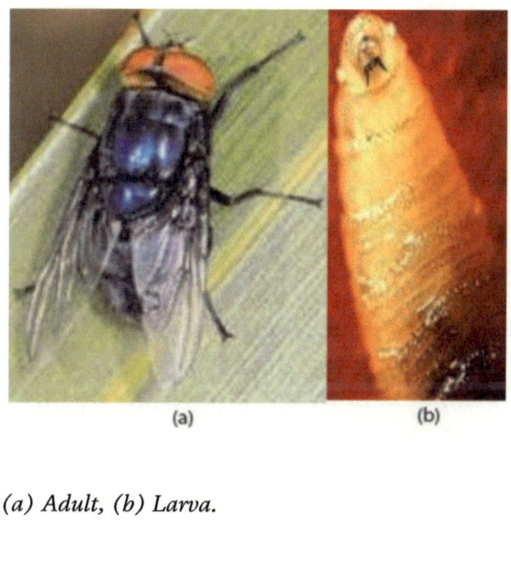

Figure 26.
Cochliomyia hominivorax. *(a) Adult, (b) Larva.*

are non-live tissue feeders and larvae mature in decomposing organic material. Identity of larvae is determined by morphologic features of the spiracle openings. In various cases, distinction from primary screwworm is of significance. Some species of blowflies are vital in occurrence of 'Limber-neck' or avian botulism. Epidemics are shared in both captive and free-ranging birds. Deceased birds aid as a food source of growing maggots that serve as resources of food and infection for other birds.

15.2 Treatments

Different treatments for controlling of invasions are available based on the dipterans circumstances and the parasite involved. Adults of Diptera are usually controlled by application of spray products such as pyrethrins and malathion, applied to premises and done directly at adult flies. Larvae of flies can be managed by different means liable to their location. Maximum myiasis creating larvae are controlled effectively with Ivermectin in the course of their migratory stage of life cycle. Many organophosphates (topical, sprays, etc.) are available as well. The programing of treatment can be imperative as killing of migrating larvae at specific time and location in the body can cause in tissue reactions and pathologic changes.

Long term control of midges and gnats requires trying to eliminate breeding sites such as wet areas or standing water. However, this type of control is not practical at large scale. Often, water should not be treated with any insecticide in an attempt to control gnats. The potential harm to the environment and wildlife is too great to justify an application for a temporary nuisance.

Because larvae require oxygen, blocking the skin opening of host may cause them to leave or at least come closer to the surface. When they are closer to the surface, it is easier to pull them out with forceps. Sometimes physicians inject an anesthetic into the skin, make a small incision and pull the larva out with forceps. The drug ivermectin, given by mouth or applied to the skin, also may kill the larva or cause it to leave host.

At certain times of the year, when livestock are most vulnerable to flies (castration, birthing, etc.,) if possible their daily inspections should be done. As with many things, prevention is the best cure and any open wound, even so small as a blister, is a potential infestation site, which should be treated accordingly with approved pesticides. In addition to the continued release of sterile males, a screwworm adult suppression system is now used, which involves a chemical attractant with dichlorvos.

16. *Dipteran* usage in industries

Insects harbor high potential for nonfood usage as antimicrobial effects, additives and even cosmetics and pharmaceuticals. Recently, farming insects have been emerged as a new source of protein and lipid production. Investigations have been performed for proteomics and lipidomics on black soldier fly *Hermetia illucens* (Linnaeus) of family *Stratiomyidae* (**Figure 27**) and blow fly (*L. sericata*) larvae. The result displayed great levels of lauric acid in soldier fly that after biological decomposition could even increase. Proteomics study exposed the presence of proteins like collagen with a cosmetic interest, and proteins with antimicrobial properties such as phenoloxidases and enzymatic actions, like trypsin and amylase. Black fly larvae can be used as fresh, frozen, freeze-dried or meals for feeding of animal proteins to fish and poultry. The substitution of 10% soybean meal by the same amount of fly larvae meal improves weight gain, conversion rate and carcass yield. Black fly can be a valuable candidate for mass rearing on agro-industrial wastes or by-products.

Figure 27.
Lucilia sericata. *(a) Adult, (b) Larvae.*

Furthermore, its rearing brings several collateral benefits, for example reducing the smell of decaying organic matter or the production of biofertilizer. Black fly larvae are capable to transform nutrients from plants, residues and other agricultural by-products into compounds that are digestible by monogastric animals. Such information exposes new opportunities for future research in cosmetic and pharmacological approaches to discover novel molecules of interests [99, 100].

17. Evolution and paleontology

Dipteran insects are endopterygotes that go through an essential metamorphosis. The ownership of a single pair of complete wings differentiates maximum of true flies from other insects having the word 'fly' in their names (whiteflies, scorpionflies, hangingflies, caddisflies and butterflies). They belong to the Mecopterida, alongside the Mecoptera. On the other hand, some true flies such as louse flies (Hippoboscidae) have been converted to secondarily wingless. The earliest fly fossils found so far are from the Triassic period [geologic period and system which spans 50.6 million years from the end of the Permian period 251.9 million years ago (Mya), to the beginning of the Jurassic period 201.3 (Mya)], about 240 million years ago. Phylogenetic analysis suggests that flies originated in the Permian period [geologic period and system, which spans 47 million years from the end of the Carboniferous period 298.9 million years ago (Mya), to the beginning of the Triassic period 251.902 (Mya)], about 260 million years ago. Diptera belongs to panorpoid complex that consists of Trichoptera (caddisflies), Mecoptera (scorpionflies), Siphonaptera (fleas), Lepidoptera (butterflies and moths) and Diptera (true flies). The whole are thought to have grown as of an ancestor, which existed in moss, and four-winged insects that look like crane flies and have been well-maintained as fossils in Permian deposits, rocks set down between 299 million and 251 million years before. Strata of the Lower Jurassic System (from about 201 million to 174 million years back) comprise several true midges. Initial Brachycera initiated to be visible in the Mesozoic Era (about 252 million–66 million years past). Cyclorrhapha seemed in the Cretaceous period (145 million–66 million years back). By the finish of the Eocene Epoch, certain 34 million years before, maximum new families of flies have been developed. Flies in copal and amber dated to the Oligocene Epoch (about 34 million–23 million years past) are related to living genera.

A determined phylogeny for flies delivers a background for developmental, genomic and evolutionary homework by facilitating assessments across model organisms. Up till now, recent research has advocated that fly relations have been out of sight by manifold episodes of fast diversification. A phylogenomic estimate of fly relations based on morphology and molecules has been delivered from 149 of 157 families, comprising 30 kb from complete mitochondrial genomes and 14 nuclear loci pooled with 371 morphological characters. Manifold studies display support for traditional groups (Brachycera, Cyclorrhapha and Schizophora) and verify contentious discoveries, for instance, the anomalous Deuterophlebiidae as the sister cluster to entire remaining Diptera. Conclusions disclose that the closest lineages of the Drosophilidae are much adjusted parasites (including the wingless Braulidae) of bees and other insects. Moreover, micro-RNAs have been used to decide a node with suggestions for evolution of embryonic development in Diptera. It has been confirmed that flies practiced three episodes of quick radiation of lower Diptera (220 Mya), lower Brachycera (180 Mya) and Schizophora (65 Mya), and a number of life history changeovers to phytophagy, hematophagy and parasitism in the history of fly evolution above 260 million years [101].

18. Conclusion

Flies are one of four super radiations of insects (along with beetles, wasps and moths) that account for the majority of animals life on earth. This is one of the largest insect orders in the world and includes many familiar insects such as mosquitoes, midges, sand flies, house flies and blow flies. This handsome book chapter definitive works on creatures of the order Diptera, by combining scholarly thoroughness and new perspective on descriptions, diversity, life histories, behavior, classifying and identifying, interactions with plants and animals, origins and distribution, transmitting diseases, pollinating plants, disposing of dung and carrion, natural life, and gives advice on how to control them as well as a detailed global overview of fly families and subfamilies on the planet. Diptera can be distinguished by the features like one pair of membranous wings, hind wings are reduced to small club like structures called halteres used as stabilizers during flight, sucking mouthparts, large compound eyes and short simple antennae, frilled or bushy in mosquitoes and crane flies. Flies mate while flying, eggs are usually laid on an appropriate food source, larvae complete development where these are laid and pupate in the substrate, which may be soil, plant tissue or animal tissue, organic matter and water. Owing to sucking and piercing mouthparts, adult flies are able to only ingest liquid foods, mostly digestion is to some extent external and salivary secretions are presented to liquefy diet and then softened product is consumed up. Some march flies and mosquitoes with their proboscis, pierce skin of prey and then suck up blood. Larvae of these insects order generally feed on decomposing moist food things such as fungi, carrion, rotting vegetable matter and dung, while some are parasites or predators of other animals. Partakers of this order have the greatest diversity of species and are found in almost all types of terrestrial and freshwater habitats. Diptera includes species known for their ubiquity (*M. domestica* house fly), their role as pests (*Anopheles gambiae* Giles malaria mosquito) and their value as model organisms across the biological sciences (*Drosophila melanogaster* Meigen). Many species of Diptera are important due to the role they play in disease transmission; such as biting midge (no-see-ums, sand flies) attacks in areas of low light without wind; gnats attack at hairline and are active during moist times (early spring); black fly bites during day near streams, is attracted to dark moving objects) and disease vector for onchocerciasis; horse fly, deer fly or gad fly bite on warm, cloudy days; sand fly (small, moth-like insect) bites at night in damp areas and is disease vector for leishmaniasis; snipe fly bites during daytime and is risk of severe allergic reaction; stable fly bites during daytime (especially during thunder storms); and tsetse fly bites during daytime even through clothing and is disease vector for African trypanosomiasis. In practice, protection of communities and animals is sometimes possible by avoiding places where mosquitos and biting flies are known to rest or breed, and by not visiting risky places during peak biting hours could be an effective way to reduce their exposure to biting insects and the transmission of diseases.

Author details

Muhammad Sarwar
National Institute for Biotechnology and Genetic Engineering (NIBGE), Faisalabad, Pakistan

*Address all correspondence to: drmsarwar64@gmail.com

References

[1] Capinera JL. Encyclopedia of Entomology. 2nd ed. Netherlands: Springer Science & Business Media; 2008. p. 1481

[2] Drake CM. Provisional Atlas of the Larger Brachycera (Diptera) of Britain and Ireland. Huntingdon, UK: Biological Records Centre, Institute of Terrestrial Ecology; 1991. p. 131

[3] Sarwar M. Dissemination of infectious agents of human diseases via insects vectors of public health prominence. American Journal of Clinical Neurology and Neurosurgery. 2015;**1**(3):169-174

[4] Colless DH, McAlpine DK. Diptera (flies). In: The Division of Entomology. Commonwealth Scientific and Industrial Research Organisation, Canberra (spons.), The Insects of Australia. Melbourne: Melbourne Univ. Press; 1991. pp. 717-786

[5] Skevington JH, Dang PT. Exploring the diversity of flies (Diptera). Biodiversity. 2002;**3**:3-27

[6] Mogi M, Chan KL. Predatory habits of dipteran larvae inhabiting *Nepenthes* pitchers. The Raffles Bulletin of Zoology. 1996;**44**(1):233-245

[7] Courtney GW, Pape T, Skevington JH, Sinclair BJ. Biodiversity of Diptera. In: Foottit R, Adler P, editors. Insect Biodiversity: Science and Society. 1st ed. USA: Blackwell Publishing; 2009. pp. 185-187, 222

[8] Sarwar M. Insect vectors involving in mechanical transmission of human pathogens for serious diseases. International Journal of Bioinformatics and Biomedical Engineering. 2015;**1**(3): 300-306

[9] Sarwar M. Direct possessions of insect arthropods on humans owing to allergen, blood sucking, biting, envenomation and stinging side by side case diagnosis and treating. International Journal of Bioinformatics and Biomedical Engineering. 2015;**1**(3): 331-337

[10] Sarwar M. Insect borne diseases transmitted by some important vectors of class insecta hurtling public health. International Journal of Bioinformatics and Biomedical Engineerin. 2015;**1**(3): 311-317

[11] Sarwar M. Occurrence of insect pests on guava (*Psidium guajava*) tree. Pakistan Journal of Zoology. 2006; **38**(3):197-200

[12] Sarwar M. Problem created owing to insects in carrying vector borne diseases and combined vector control approach. International Journal of Chemical and Biomolecular Science. 2015;**1**(4): 303-309

[13] Sarwar M. Management of guava (*Psidium guajava*) orchard against insect pests. Economic Review. 2006;**8/9**(38): 28-30

[14] Sarwar M. Incidence of insect pests on ber (*Zizyphus jujube*) tree. Pakistan Journal of Zoology. 2006;**38**(4):261-263

[15] Ssymank A, Kearns CA, Pape T, Thompson FC. Pollinating flies (Diptera): A major contribution to plant diversity and agricultural production. Biodiversity. 2008;**9**(1–2):86-89

[16] Riaz M, Sarwar M. A new record of fruit fly *Trupanea amoena* (Frauenfeld) within genus *Trupanea* Schrank of subfamily Tephritinae (Diptera: Tephritidae) from Pakistan. Journal of Zoological Sciences. 2013;**1**(2):7-12

[17] Riaz M, Sarwar M. Fruit fly *Tephritis zernyi* Hendel, a member of genus

Tephritis Latreille in family Tephritidae within order Diptera. Journal of Agriculture and Allied Sciences. 2014; **3**(4):17-23

[18] Riaz M, Sarwar M. A new record of safflower fly *Acanthiophilus helianthi* (Rossi) of genus *Acanthiophilus* Becker in subfamily Tephritinae (Diptera: Tephritidae) from the Fauna of Pakistan. Journal of Agriculture and Allied Sciences. 2014;**3**(1):39-44

[19] Sarwar M, Riaz M. New distribution records of fruit fly *Dacus sphaeroidalis* (Bezzi) (Diptera: Tephritidae) From Pakistan and improved description of this pest species. Journal of Zoological Sciences. 2014;**2**(1):1-6

[20] Sarwar M, Hamed M, Rasool B, Yousaf M, Hussain M. Host preference and performance of fruit flies *Bactrocera zonata* (Saunders) and *Bactrocera cucurbitae* (Coquillett) (Diptera: Tephritidae) for various fruits and vegetables. International Journal of Scientific Research in Environmental Sciences. 2013;**1**(8):188-194

[21] Sarwar M, Hamed M, Yousaf M, Hussain M. Monitoring of population density and fruit infestation intensity of tephritid fruit flies (Diptera: Tephritidae) in *Citrus reticulata* Blanco orchard. Journal of Zoological Sciences. 2014;**2**(3):1-5

[22] Sarwar M, Hamed M, Yousaf M, Hussain M. Surveillance on population dynamics and fruits infestation of tephritid fruit flies (Diptera: Tephritidae) in mango (*Mangifera indica* L.) orchards of Faisalabad, Pakistan. International Journal of Scientific Research in Environmental Sciences. 2014;**2**(4):113-119

[23] Sarwar M, Hamed M, Yousaf M, Hussain M. Monitoring of population dynamics and fruits infestation of tephritid fruit flies (Diptera: Tephritidae) in guava (*Psidium guajava* L.) orchard. Journal of Agriculture and Allied Sciences. 2014;**3**(2):36-40

[24] Sarwar M. Management of banana (*Musa paradisiaca* Linnaeus) orchard against insect pests. Federal Urdu University of Arts, Science and Technology Journal of Biology. 2011; **1**(1):107-110

[25] Sarwar M. Quarantine treatments for mortality of eggs and larvae of fruit flies (Diptera: Tephritidae) invading fresh horticulture perishable produces. International Journal of Animal Biology. 2015;**1**(5):196-201

[26] Sarwar M. Cultural measures as management option against fruit flies pest (Tephritidae: Diptera) in garden or farm and territories. International Journal of Animal Biology. 2015;**1**(5): 165-171

[27] Sarwar M. Mechanical control prospectus to aid in management of fruit flies and correlated tephritid (Diptera: Tephritidae) pests. International Journal of Animal Biology. 2015;**1**(5):190-195

[28] Sarwar M. How to manage fruit fly (Family Tephritidae) pests damage on different plant host species by take up of physical control measures. International Journal of Animal Biology. 2015;**1**(4): 124-129

[29] Sarwar M. Biological control program to manage fruit fly pests and related tephritids (Diptera: Tephritidae) in backyard, landscape and garden. International Journal of Animal Biology. 2015;**1**(4):118-123

[30] Sarwar M. Birth control for insects: The Sterile Insect Technique (SIT) for controlling fruit fly (Tephritidae: Diptera) by releasing sterile males. International Journal of Animal Biology. 2015;**1**(5):253-259

[31] Sarwar M. Attraction of female and male fruit flies (Diptera: Tephritidae) to

bait spray applications for reduction of pest populations. International Journal of Animal Biology. 2015;**1**(5):225-230

[32] Sarwar M. The role of male annihilation technique to get rid of notorious fruit flies (Tephritidae: Diptera) in fruit and vegetable farms. International Journal of Animal Biology. 2015;**1**(5):260-265

[33] Sarwar M. Field tests for exploiting the behavioral control tactics to pest tephritid fruit flies (Insecta: Diptera). International Journal of Animal Biology. 2015;**1**(5):243-248

[34] Sarwar M. Genetic control tactic against fruit flies (Diptera: Tephritidae) insect to escape destruction of perishable horticulture crops. International Journal of Animal Biology. 2015;**1**(5):209-214

[35] Sarwar M. Area-wide integrated management of fruit flies (Diptera: Tephritidae) pest in vegetables cultivation. Journal of Biological and Environmental Engineering. 2016;**1**(2): 10-16

[36] Sarwar M. An area-wide integrated management of fruit fly (Diptera: Tephritidae) pests in fruits production. International Journal of Plant Science and Ecology. 2018;**4**(1):1-7

[37] Sarwar M, Ahmad N, Rashid A, Shah SMM. Valuation of gamma irradiation for proficient production of parasitoids (Hymenoptera: Chalcididae & Eucoilidae) in the management of the peach fruit-fly, *Bactrocera zonata* (Saunders). International Journal of Pest Management. 2015;**61**(2): 126-134

[38] Shah SMM, Ahmad N, Sarwar M, Tofique M. Rearing of *Bactrocera zonata* (Diptera: Tephritidae) for parasitoids production and managing techniques for fruit flies in mango orchards. International Journal of Tropical Insect Science. 2014;**34**(1):108-113

[39] Sarwar M. Life history of house fly *Musca domestica* Linnaeus (Diptera: Muscidae), its involvement in diseases spread and prevention of vector. International Journal for Research in Applied Chemistry. 2016;**1**(7):23-34

[40] Morales GE, Wolff M. Insects associated with the composting process of solid urban waste separated at the source. Revista Brasileira de Entomologia. 2010;**54**(4):645-653

[41] Smith KGV. An introduction to the immature stages of British flies. Diptera Larvae, with notes on eggs, puparia and pupae. Handbooks for the Identification of British Insects. 1989; **10**(14):1-280

[42] Sarwar M. Skin disorders inflicted through insect invertebrates along with diagnosis and treating of cases. Journal of Nanoscience and Nanoengineering. 2015;**1**(4):233-240

[43] McAlpine JF. Morphology and terminology. In: McAlpine JP et al., editors. Manual of Nearctic Diptera. Vol. 1. Ottawa: Research Branch, Agriculture Canada; 1981. pp. 9-63. Monograph 27

[44] Bernhard M, Haenni JP. Morphology and terminology of adult Diptera (other than terminalia). In: Papp L, editor. Bela Darvas Contributions to a Manual of Palaearctic Diptera. General and Applied Dipterology. Vol. 1. Budapest: Science Herald; 2000. pp. 22-51

[45] Sarwar M. Foodstuff contaminations with foodborne pathogens vehicled by insect vectors. International Journal of Bioinformatics and Biomedical Engineering. 2015;**1**(3): 352-358

[46] Sarwar S, Sarwar M. Involvement of insects (Insecta: Artropoda) in spreading of plant pathogens and approaches for pests management.

American Journal of Microbiology and Immunology. 2018;**3**(1):1-8

[47] Thorp JH, Rogers DC. Field Guide to Freshwater Invertebrates of North America. 1st ed. London, UK: Academic Press; 2010. p. 304

[48] Merritt RW, Wallace JR. The role of aquatic insects in forensic investigations. In: Byrd JH, Castner JL, editors. Forensic Entomology: The Utility of Arthropods in Legal Investigations. Boca Raton, FL, USA: CRC Press; 2001. pp. 177-222

[49] Tabor KL, Fell RD, Brewster CC. Insect fauna visiting carrion in Southwest Virginia. Forensic Science International. 2005;**150**(1):73-80

[50] O'Flynn MA. The succession and rate of development of blowflies in carrion in southern Queensland and the application of these data to forensic entomology. Journal of the Australian Entomological Society. 1983;**22**:137-148

[51] Anderson GS. Minimum and maximum development rates of some forensically important Calliphoridae (Diptera). Journal of Forensic Sciences. 2000;**45**:824-832

[52] Courtney GW, Cranston PS, Order Diptera. In: Thorp JH, Rogers DC, editors. Thorp and Covich's Freshwater Invertebrates. 4th ed. London, UK: Academic Press; 2015. pp. 1043-1058

[53] Vockeroth JR, Thompson FC. Syrphidae. In: McAlpine JF, Peterson BV, Shewell GE, Teskey HJ, Vockeroth JR, Wood DM, editors. Manual of Nearctic Diptera. Ottawa: Canadian Government Publishing Centre; 1987. pp. 713-743

[54] Mayhew PJ. Why are there so many insect species? Perspectives from fossils and phylogenies. Biological Reviews. 2007;**82**(3):425-454

[55] Rohdendorf BB. Historical Development of Dipterous Insects. Vol. 100. Moscow: Transactions of the Institute Paleontology, Academy of Sciences. USSR; 1964. p. 311

[56] Griffiths GCD. Relationships among the major subgroups of Brachycera (Diptera): A critical review. Canadian Entomologist. 1994;**126**:861-880

[57] Oosterbroek P, Courtney G. Phylogeny of the nematocerous families of Diptera (Insecta). Zoological Journal of the Linnean Society. 1995;**115**:267-311

[58] Michelsen V. Neodiptera: New insights into the adult morphology and higher level phylogeny of Diptera (Insecta). Zoological Journal of the Linnean Society. 1996;**117**:71-102

[59] Delfinado MD, Hardy DE. A Catalog of the Diptera of the Oriental Region: Suborder Nematocera. Vol. 1. Honolulu: University Press of Hawaii; 1973. p. 618

[60] Delfinado MD, Hardy DE. A Catalog of the Diptera of the Oriental Region: Suborder Brachycera Through Division Aschiza, Suborder Cyclorrhapha. Vol. 2. Honolulu: University Press of Hawaii; 1975. p. 459

[61] Delfinado MD, Hardy DE. A Catalog of the Diptera of the Oriental Region. Suborder Cyclorrhapha (Excluding Division Aschiza). Vol. 3. Honolulu: University Press of Hawaii; 1977. p. 854

[62] Oosterbroek P. The European Families of the Diptera Identification, Diagnosis, Biology. Netherlands: Royal Dutch Society for Natural History (KNNV) Publishing; 2015. p. 208

[63] Brown BV, Borkent A, Cumming JM, Wood DM, Woodley NE, Zumbado M. Manual of Central American Diptera. Vol. 1. Ottawa: NRC Research Press; 2009. p. 714

[96] Cogley TP, Cogley MC. Field observations of the host-parasite relationship associated with the common horse bot fly, *Gasterophilus intestinalis*. Veterinary Parasitology. 2000;**88**:93-105

[97] Ockenhouse CF, Samlaska CP, Benson PM, Roberts LW, Eliasson A, Malane S, et al. Cutaneous myiasis caused by the African tumbu fly (*Cordylobia anthropophaga*). Archives of Dermatology. 1990;**126**(2):199-202

[98] Sherman RA, Hall MJR, Thomas S. Medicinal maggots: An ancient remedy for some contemporary afflictions. Annual Review of Entomology. 2000; **45**:55-81

[99] Rabani V, Cheatsazan H, Davani S. Proteomics and lipidomics of black soldier fly (Diptera: Stratiomyidae) and blow fly (Diptera: Calliphoridae) larvae. Journal of Insect Science. 2019;**19**(3):29

[100] Caruso D, Devic E, Subamia IW, Talamond P, Baras E. Technical Handbook of Domestication and Production of Diptera Black Soldier Fly (BSF) *Hermetia illucens*, Stratiomyidae. Bogor, West Java, Indonesia: PT Penerbit IPB Press; 2013. p. 141

[101] Wiegmann BM, Trautwein MD, Winkler IS, Barr NB, Kim JW, Lambkin C, et al. Episodic radiations in the fly tree of life. In: Proceedings of the National Academy of Sciences (PNAS). 2011;**108**(14):5690-5695

Chapter 2

Taxonomic Shifts in *Philornis* Larval Behaviour and Rapid Changes in *Philornis downsi* Dodge & Aitken (Diptera: Muscidae): An Invasive Avian Parasite on the Galápagos Islands

Lauren K. Common, Rachael Y. Dudaniec, Diane Colombelli-Négrel and Sonia Kleindorfer

Abstract

The parasitic larvae of *Philornis* downsi Dodge & Aitken (Diptera: Muscidae) were first discovered in Darwin's finch nests on the Galápagos Islands in 1997. Larvae of *P. downsi* consume the blood and tissue of developing birds, causing high in-nest mortality in their Galápagos hosts. The fly has been spreading across the archipelago and is considered the biggest threat to the survival of Galápagos land birds. Here, we review (1) *Philornis* systematics and taxonomy, (2) discuss shifts in feeding habits across *Philornis* species comparing basal to more recently evolved groups, (3) report on differences in the ontogeny of wild and captive *P. downsi* larvae, (4) describe what is known about adult *P. downsi* behaviour, and (5) discuss changes in *P. downsi* behaviour since its discovery on the Galápagos Islands. From 1997 to 2010, *P. downsi* larvae have been rarely detected in Darwin's finch nests with eggs. Since 2012, *P. downsi* larvae have regularly been found in the nests of incubating Darwin's finches. Exploring *P. downsi* ontogeny and behaviour in the larger context of taxonomic relationships provides clues about the breadth of behavioural flexibility that may facilitate successful colonisation.

Keywords: *Protocalliphora*, *Passeromyia*, *Philornis*, nest larvae, hematophagous, subcutaneous, Darwin's finches, Passeriformes

1. Introduction

Three genera of flies within the order Diptera have larvae that parasitise avian hosts: *Protocalliphora* Hough (Calliphoridae), as well as *Passeromyia* Rodhain & Villeneuve (Muscidae) and *Philornis* Meinert (Muscidae). The adult flies in these genera are free-living and do not parasitise birds, but their larvae develop in the nests of altricial birds, feed on their avian hosts, and exhibit feeding behaviours from hematophagy to coprophagy [1, 2]. Most larval infestations have been documented in host nests of the order Passeriformes, but larvae have also been found in nests of Accipitriformes, Apodiformes, Strigiformes and other avian taxa (*Protocalliphora*: [3]; *Passeromyia*: [4]; *Philornis*: [5, 6]). The effect of these parasitic fly larvae on host survival can be severe to mild, depending on many factors including host population size, body size, nesting density and the presence of behavioural or immunological defence mechanisms [6–8].

Protocalliphora is widely distributed throughout the Holarctic and contains 40+ species with obligate avian parasitic larvae [3]. Within Muscidae, only *Passeromyia* and *Philornis* larvae parasitise birds [4, 9, 10]. Both *Passeromyia* and *Philornis* are members of the subfamily Cyrtoneurininae, however their complete evolutionary relationships have yet to be resolved [11, 12]. Due to the similarities between *Passeromyia* and *Philornis*, many workers regarded the two genera as close relatives, including Skidmore [9], who stated that their similarities could not be based on convergent evolution alone. The five *Passeromyia* species include *P. steini* (Pont), *P. heterochaeta* (Villeneuve), *P. indecora* (Walker), *P. longicornis* (Macquart) and *P. veitchi* (Bezzi), and are distributed throughout Europe, Africa, Asia and Australasia [4, 13]. *Passeromyia* species differ in their larval habits. For example, *P. steini* larvae scavenge nests for organic matter and *P. indecora* larvae consume host resources as subcutaneous parasites. The 52 *Philornis* species are distributed primarily in Neotropical South America and southern North America [1, 2, 10]. *Philornis* species also show a wide range of feeding habits, including free-living coprophagous larvae, free-living semi-hematophagous larvae, and subcutaneous hematophagous larvae (**Table 1**). One species, *P. downsi*, is a recently discovered invasive species on the Galápagos Islands [14, 15]. Its semi-hematophagous larvae cause significant in-nest host mortality in their novel Galápagos land bird hosts [16]. Cladistics and molecular phylogenetic analyses suggest that the parasitic larval habits of *Passeromyia* and *Philornis* evolved independently [10, 12] despite the similarities between both genera including cocoon-wrapped puparia, life history, and clade.

The Galápagos Islands have been listed as a World Heritage site in 1978. Given a suite of threats, including introduced species, the archipelago was added to the 'List of the World Heritage in Danger' in 2007 and then removed from this list in 2010 because of actions by the Government of Ecuador to reduce invasions [17, 18]. Biological invasion is considered the greatest threat to biodiversity in the Galápagos Islands [19]. Currently, 543 terrestrial species have been introduced, of which 55 are considered harmful or potentially harmful to native biodiversity [17].

In this chapter, we consider changes in the development and behaviour of the accidentally introduced fly *P. downsi* Dodge and Aitken (Diptera: Muscidae), that is now considered the biggest threat to the survival of Galápagos land birds [20]. The first *P. downsi* larvae were collected from Galápagos land bird nests on Santa Cruz Island in 1997 [21]. From examination of museum specimens collected in 1899 (during the Stanford University Expedition led by Robert Snodgrass and Edmund Heller), in 1905–1906 and 1932 (during expeditions sponsored by the California Academy of Sciences), and in 1962 (by Robert Bowman) on Floreana Island, there is no current evidence to suggest *P. downsi* was present or abundant on the Galápagos Islands prior to 1964, though this is possible [22, 23]. By collating information from a range of researchers investigating *Philornis* in general and *P. downsi* in particular, we aim to improve our understanding of the ontogeny and behaviour of an invasive *Philornis* species within the larger context of Dipteran parasites of birds. We review *Philornis* systematics and taxonomy, discuss feeding habits across *Philornis* species, report on differences in the ontogeny of wild and captive *P. downsi* larvae, report on adult *P. downsi* behaviour, and describe changes in *P. downsi* behaviour since its discovery on the Galápagos Islands.

***Aitkeni* group**	**Larval habits**	***Falsificus* group**	**Larval habits**	***Angustifrons* group**	**Larval habits**
P. fasciventris [37]	FLC	*P. fumicosta*		*P. downsi* [30]	FLSH
P. schildi		*P. univittatus*		*P. niger* [1, 30]	SubH
P. amazonensis		*P. grandis*		*P. porteri*[1] [43]	SubH
P. lopesi		*P. sabroskyi*		*P. mimicola*[2] [40]	SubH
P. aikteni [30]	FLC	*P. falsificus* [1, 30]	FLSH	*P. sperophilus* [1]	SubH
P. zeteki				*P. carinatus* [47]	SubH
P. rufoscutellaris [36]	FLC			*P. deceptiva* [48, 49]	SubH
P. rettenmeyeri				*P. trinitensis* [30]	SubH
P. setinervis				*P. glaucinis* [30]	SubH
				P. pici [24]	SubH
				P. vespidicola[3] [2]	SubH
				P. medianus [33]	SubH
				P. vulgaris [1]	SubH
				P. masoni	SubH
				P. diminutus [1]	SubH
				P. querulus [30]	SubH
				P. albuquerquei	
				P. frontalis [1]	SubH
				P. gagnei [33]	SubH
				P. insularis [33]	SubH
				P. obscurinervis	
				P. petersoni	
				P. torquans [1]	SubH
				P. angustifrons [30]	SubH
				P. bellus [2]	SubH
				P. sanguinis [30]	SubH
				P. seguyi[4] [50, 51]	SubH

[1]*Some P. porteri larvae found in ear canals and nares of nestlings; some later instars found feeding externally on abdomen and wings [41, 43].*
[2]*P. mimicola larvae found in the nasal cavity of owls, mainly subcutaneous on body [40].*
[3]*Only known specimens of P. vespidicola collected from nests of the wasp Paracharitopus frontalis (Hymenoptera: Vespidae) [2, 29].*
[4]*P. nielseni proposed synonym of P. seguyi [34].*

Table 1.
Philornis *species ordered according to taxonomy, from the most basal* 'aitkeni-*group' to the most recently evolved* 'angustifrons-*group' (groups from [33]). Larval feeding habits are shown where known and abbreviated as follows: free-living coprophagous larvae (FLC), free-living semi-hematophagous larvae (FLSH), subcutaneous hematophagous larvae (SubH). The following nine species are not included in the list as they are currently not assigned to a taxonomic group [33] given insufficient information: P. molesta, P. nielseni, P. blanchardi, P. cinnamomina, P. convexus, P. mima, P. obscurus, P. steini, P. umanani.*

2. *Philornis* systematics and taxonomy

Macquart [24] provided the first description of *Philornis* larvae when he described *Aricia pici*; a subcutaneous larval parasite found on an adult Hispaniolan woodpecker (*Melanerpes striatus;* previously *Picus striatus*) Muller (Piciformes: Picidae). Meinert [25] erected the genus *Philornis* for the single species, *P. molesta*,

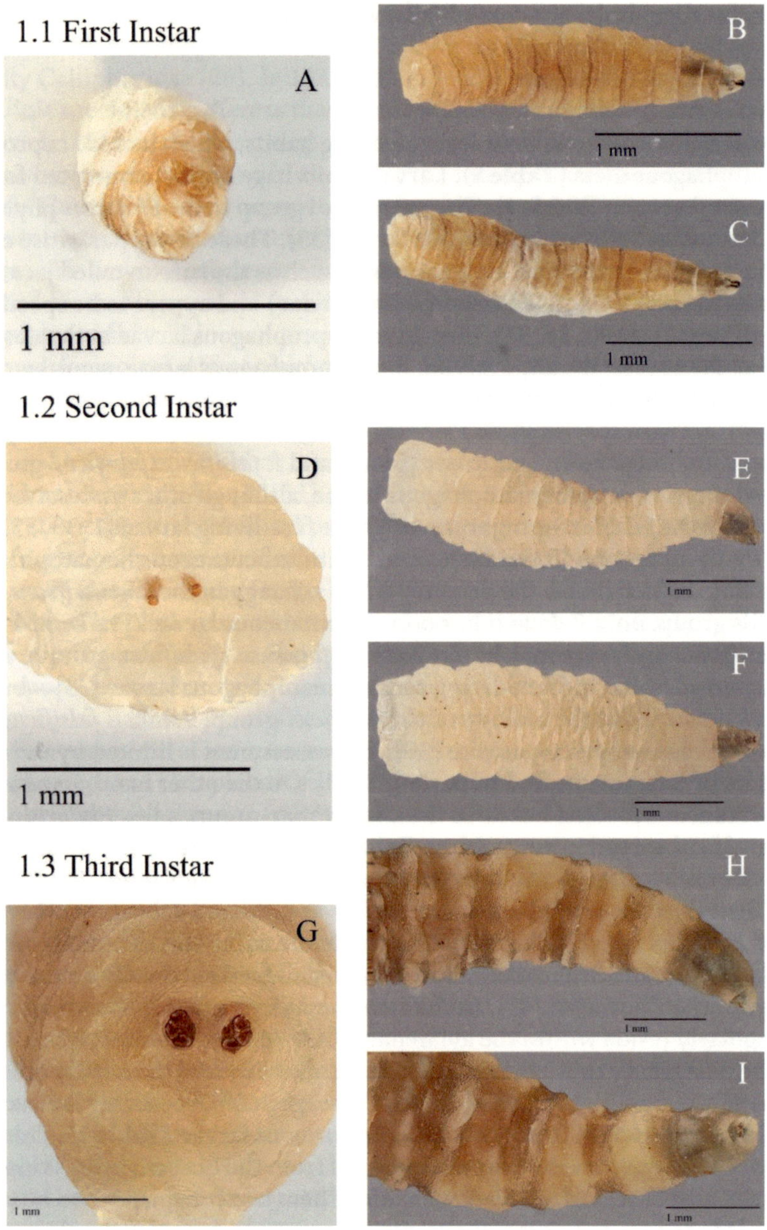

Figure 1.
Three larval stages of Philornis downsi. (1) First instar: (A) posterior spiracles, (B) lateral view, (C) ventral view. (2) second instar: (D) posterior spiracles, (E) lateral view, (F) ventral view. (3) third instar: (G) posterior spiracles, (H) lateral view, (I) ventral view. Obtained by the authors from larvae collected on Floreana Island, Galápagos, Ecuador between 2010 and 2014. The photographs were taken using a visionary digital LK imaging system (dun, Inc) with a canon EOS 5DsR camera and capture one pro 11.3.1, phase one (Flinders University). Images were produced using Zerene stacker 1.04, Zerene systems LLC, software, and cropped and resized in Photoshop CS5.

on the blood and fluids of their host by penetrating the skin of the nestlings [2, 30]. Larval instar morphological descriptions are given by Fessl et al. [44]. The most distinct character between the instars is the posterior spiracles, which change in colour, shape and number of spiracular slits present throughout larval development [44].

Figure 1(1A) shows the posterior spiracles of a 1st instar *P. downsi* larva, characterised by their light pigmentation and two oval slits present [44]. The spiracles of a

1st instar larva are separated by slightly more than their diameter. First instar larvae lack anterior spiracles (**Figure 1(1B)**). The posterior spiracles of a 2nd instar larva are similarly round with two oval slits; however, the distance between them is two to three times of their diameter (**Figure 1(2D)**; [44]). Anterior spiracles are present during the 2nd instar, and semicircular in shape, lightly pigmented and visible in (**Figure 1(2E)**). 3rd instar posterior spiracular plates are darkly pigmented and round in shape, distinct C-shaped spiracular slits radiate from median ecdysial scar (**Figure 1(3G)**). Pigmentation of the median ecdysial scar is light in early 3rd instar larvae and becoming darkly pigmented later in the stage. Semi-circular anterior spiracles are retained in 3rd instar larvae (**Figure 1(3H)**). Cephaloskeleton morphology differs between instars as outlined in Fessl et al. [44]. Recent studies report a decrease in *P. downsi* puparia size across 2004–2014 [54]. Common et al. (unpublished data), and hence body size is certainly not a useful method to classify instars. In general, it is recommended to use a suite of morphological characters, including anterior and posterior spiracular morphology, to determine the larval instar.

4.2 Larval development

The developmental period of *Philornis* larvae is associated with the species' larval feeding habit. For example, time to pupation in coprophagous species takes up to 29 days, but only 4–8 days in subcutaneous species [2, 55]. Larval development periods in free-living species such as *P. downsi* are difficult to determine in the wild as the host nest needs to be dismantled to observe the larvae. Early studies of abandoned or failed nests found 1st instar larvae in nests with 1–3 day old nestlings, 2nd instars in nests with 3–6 day old nestlings and 3rd instars in nests with 3–10 day old nestlings [44]. Larval collections following the cessation of activity at host nests suggest that the minimum time for pupation in *P. downsi* is 4–7 days [54].

Compared with larval development times in the wild, larval development times under laboratory conditions are longer. First attempts to rear *P. downsi* larvae in the absence of a living host had a low success rate, with only three larvae out of 477 reaching the adult stage after a 36 day development time (mean 18 day larval development, 12 day pupation) [56]. As the diet for rearing larvae in captivity was refined, the success rate increased to 10% and larval development time decreased [57]. Development time in the laboratory ranged from 9 to 10 days from larva to pupa [57] with even faster development times occurring as the rearing conditions have improved [pers. comm. P. Lahuatte]. Egg hatch rates in captivity have been high (96%), with most mortality in 1st instar larvae (77%) [57]. Laboratory-based diets that have been developed in the absence of a bird host are primarily based on chicken blood, with more successful diets including hydrolysed protein and vitamin fortification [57]. The lack of keratin in the diet may be causing elevated 1st instar mortality, as 1st instars consume the keratin of the beak in which they reside [44], however the true cause is unknown.

5. *Philornis downsi* adult behaviour

The behaviour of adult *P. downsi* is much less understood than that of the larvae. The adult fly is vegetarian, feeding on decaying fruits and flowers, including the invasive blackberry (*Rubus niveus*) Thunb (Rosales: Rosaceae) [9, 15, 31]. *Philornis downsi* is commonly attracted to a mix of blended papaya and sugar, which is used to trap adult flies (developed by P. Lincango and C. Causton; used by [58], Causton et al. in review). This mix is particularly attractive to adult flies due to the presence of yeast and fermentation products such as ethanol and acetic acid [59].

A one-year study on Floreana Island found that male and female *P. downsi* display dimorphic flight patterns, with females more likely to be caught in high and low traps (2 m, most common at 6–7 m), and males more likely to be caught in traps of intermediate height (4–5 m) [58]. As the pattern of male and female abundance are quadratic opposites, this has tentatively been suggested to be an advantage for females to avoid male flies, as frequent mating in other Dipterans has been found to decrease female reproductive success and lifespan [60, 61]. This flight pattern may also explain why certain host species experience higher parasite intensities, such as the medium tree finch (*Camarhynchus pauper*) Ridgway (Passeriformes: Thraupidae) that has an average nest height of 6 m, thus making it more susceptible to being encountered by female *P. downsi* [58, 62, 63]. However, the factors that cause bird species to experience differing intensities of *P. downsi* are complicated and vary between years. Comparison of flight height in *P. downsi* on different islands is needed to test the generality of this pattern, which may be influenced by average tree height and/or other ecological variables.

5.1 Mating behaviour

The mating behaviour of *Philornis* in general is not well understood, though there are some insights into *P. downsi* mating patterns. While mating has not been observed at or inside the nest, multiple *P. downsi* flies have been video recorded to enter host nests concurrently [45, 64]. Analysis of offspring genetic relatedness has provided estimates of the re-mating frequency of *P. downsi* [65]. Evidence for multiple mating by females has been frequently detected, and each larval infrapopulation (i.e., within nests) is sired by 1–5 males (average ~1.9 males per female) [65]. How *P. downsi* adults find each other to initiate mating is unknown. Pheromones for attraction and aggregation in muscid flies have been identified and studied [66–68]. Cuticular compounds show promise for determining if *P. downsi* produces pheromones, as females and mature males showed distinct cuticular profiles and females respond to chemicals produced by males [69–71]. Cuticular profiles could be developed as an attractant to capture flies in the field [20, 72].

5.2 Oviposition behaviour

Studies into oviposition in the genus *Philornis* have revealed that species spanning diverse larval feeding habits are oviparous [1, 9, 31, 73, 74]. This current view has previously been hotly debated, in part because the majority of species remain unstudied. Laboratory rearing and field observation have confirmed that *P. downsi* is oviparous [45, 56, 57, 75]. *Philornis* flies enter and oviposit in nests regardless of nesting phase or nestling age but have not been observed to enter nests abandoned by the parent birds during the incubation phase [45, 47]. From in-nest video recordings, *P. downsi* flies have been observed entering nests throughout the day, but generally during dusk between 1500 and 1800, with visiting rates peaking around 1700 [45, 64]. Visit length averaged 1.3–1.5 min and occurred most commonly when the adult host is away from the nest and completed once the adult host returned [45, 64]. Eggs have been generally deposited on nesting material and the base of the nest [45, 57], however on one occasion, eggs have been also laid directly by the naris of a nestling [45]. A genetic study of *P. downsi* larvae estimated that 1–6 adult females (average ~3 females) oviposit within a single nest, supporting previous observations of different sized larval groups within nests and suggesting repeated nest infestations throughout the nestling period [7, 65].

5.3 Effects of host species on *Philornis* behaviour and microbiome

Philornis downsi is one of the most generalist species within the genus, known to infest 38 host species across avian taxa [5, 6, 76]. However, this high host number may reflect the large number of studies focused on *P. downsi* due to its invasive status on the Galápagos Islands [15, 16].

It is currently unclear how *Philornis* species in general or *P. downsi* in particular find their hosts. Preliminary studies into the role of semiochemicals and volatiles in host nests as an attractant for *P. downsi* have produced inconclusive results [70]. Long-term ornithological field studies have provided some hints that the intensity of host cues may be relevant for *P. downsi* search behaviour, or alternatively that the density of host nests influences *P. downsi* oviposition behaviour. Aggregated host nests may attract *P. downsi* females due to an increase in olfactory or visual cues. These aggregated nests also provide a greater opportunity for *P. downsi* females to infest multiple nests. Indeed, small tree finch nests (*Camarhynchus parvulus*) Gould (Passeriformes: Thraupidae) with close neighbours contained more *P. downsi* larvae compared to solitary, more isolated nests [16]. Nests in areas of lower nesting density (i.e., lowlands) have been more likely to contain the offspring of a single *P. downsi* female than nests in areas of higher nesting density (i.e., highlands) that are more likely to contain the offspring of many *P. downsi* females [65]. Video recordings of adult *P. downsi* have been made inside the nests of the small ground finch (*Geospiza fuliginosa*) Gould (Passeriformes: Thraupidae), medium ground finch (*G. fortis*) Gould (Passeriformes: Thraupidae), small tree finch (*C. parvulus*) and Galápagos flycatcher (*Myiarchus magnirostris*) Gould (Passeriformes: Tyrannidae) [45, 64] (Pike et al. in prep). However, despite a combination of video recorders inside or outside the nest across studies, the recordings did not reveal information about *P. downsi* search behaviour from its flight behaviour.

A metagenomic study into *P. downsi* larval microbiome sampled from different host species found an effect of host diet on the gut bacterial community of *P. downsi* larvae [77]. Larvae retrieved from strictly insectivorous warbler finch (*Certhidea olivacea*) Gould (Passeriformes: Thraupidae) nests have a different microbiome structure compared with larvae parasitising hosts with broader dietary preferences (ground and tree finches, *Geospiza* and *Camarhynchus* sp., respectively) [77]. The gut microbiome also differed between *P. downsi* larvae (blood diet) and adults (plant diet), supporting the hypothesis that *P. downsi* microbiome changes during development and according to diet [77]. Further behavioural, biochemical and genetic studies are needed to understand *P. downsi* oviposition across host species, host locating behaviour and host specificity.

6. Changes in *P. downsi* behaviour since colonising the Galápagos Islands

6.1 Age of larval cohort in host nests

There is evidence that the oviposition behaviour of female *P. downsi* has changed since its discovery on the Galápagos archipelago. *Philornis downsi* flies are now known to oviposit during any stage of the nesting cycle [45]. In the first decades following initial discovery of *P. downsi* in Darwin's finch nests, changes in the proportions of instar classes among *P. downsi* have been observed, with evidence that oviposition occurred earlier and more synchronously in the nesting phase in the later years of the study [54]. Synchronisation in oviposition date may lead to an increase in larval competition for host resources, and as a consequence result in increased virulence for nestlings that must contend with a greater number of large,

Ref #	Year (s) of study	Island	Host species	Total no. of nests examined/no. inspected during egg phase	*P. downsi* larvae during the egg phase	Comments
[97]	2009	SC	MG	61/na		
[88]	2010	SC	MG	43/na		Female MG in parasitised nests had more *P. downsi* antibodies and spent more time standing upright when brooding than non-parasitised nests
[98]	2010	SC	MG	30/na		
[63]	2005–2010	FL	ST, MT	43/na		
[54]	2004, 2006, 2008, 2010, 2012, 2013	FL	ST, MT, SG	561/na		Evidence that *P. downsi* oviposition behaviour occurred more synchronously and earlier in nesting phase in later years of the study
[99]	2013	SC	ST, SG, MG, VGF	26/na		
[46]	2010	FL	SG	14/na		
[57]	2014	SC	GF	1/na		
[100]	2012, 2013	SC	MG, GM	127/na		
[58]	2004, 2005, 2006, 2008, 2010, 2012, 2013, 2014	FL	ST, MT, SG	254/na		
[101]	2013, 2014	SC	VGF	11/na		
[64]	2015	SC	GF	2/na		
[102]	2013	SC	MG, GM	37/na		
[86]	2012, 2014, 2015, 2016, 2017	SC	ST, WF	850/177	Yes	Larvae and puparia found in 18/72 ST nests and 52/105 WF nests that failed during egg phase; range in prevalence across species and years was 0–80% of nests
[103]	2012, 2013, 2015, 2016	SC	GM	131/na		
[104]	2010, 2013, 2014	FL	ST, MT	27/na		

The islands are abbreviated as Santa Cruz (SC), Floreana (FL), Isabela (IS), Daphne Major (DMj). The 'total number of nests examined' refers to all active nests monitored over the course of the study and 'number inspected during egg phase' is the sample size for the sub-set of nests examined during host incubation (usually following abandonment or predation) where 'na' denotes that nests have been not sampled during the egg phase. The column 'P. downsi larvae during the egg phase' states 'yes/no' referring only to nest inspections that occurred during the egg phase. Host species are abbreviated as small tree finch (ST), large tree finch (Camarhynchus psittacula) (LT), small ground finch (SG), medium ground finch (MG), woodpecker finch (Cactospiza pallida) (WP), warbler finch (Certhidea olivacea) (WF), cactus finch (Geospiza scandens) (CF), Galápagos mockingbird (GM), smooth billed ani (Crotophaga ani) (SBA), yellow warbler (Dendroica petechia) (YW), dark billed cuckoo (Coccyzus melacoryphus) (DBC), vermillion flycatcher (Pyrocephalus rubinus) (VF), vegetarian finch (Platyspiza crassirostris) (VGF), and Galápagos flycatcher (Myiarchus magnirostris) (GF).

Table 2.
Evidence of Philornis downsi larvae present in nests during incubation and before nestling hatching in studies on the Galápagos Islands.

mortality mirror that of the *P. downsi* invasion in the Galápagos Islands [6, 48, 49]. *Philornis* consumption of attending adult hosts may be an oviposition tactic that is more prevalent under conditions of resource limitation. Resource limitation could be influenced by resource termination such as early host death, resource availability when there is a limited supply of host nests (e.g., during drought), and resource accessibility, for example when competition within and between fly cohorts changes [54].

7. Conclusions

As one of three avian nest parasitic genera in Diptera, the genus *Philornis* provides a useful system to explore shifts in larval feeding behaviour in native and invasive species. *Philornis downsi* has been accidentally introduced to the Galápagos Islands and first observed in the nests of Galápagos land birds in 1997. In this chapter, we explored similarities and differences between *P. downsi* larval development and behaviour with what is known from the other 52 *Philornis* species. More basal *Philornis* (*aitkeni*-group) species have free-living coprophagous larvae and more recently evolved *Philornis* (*angustifrons*-group) tend to have subcutaneous hematophagous larvae with the exception of *P. downsi* that has free-living semi-hematophagous larvae. Since its introduction to the Galápagos Islands, there have been documented changes in the behaviour of *P. downsi*. During the early years after initial discovery of *P. downsi* on the Galápagos Islands, oviposition behaviour was asynchronous across the nesting cycle and larvae appeared to have fed exclusively on developing nestlings until 2005. In later years, *P. downsi* oviposition behaviour was earlier in the nesting cycle and more synchronous, and since 2006, larvae have also been recorded to feed on incubating females. The first records of *P. downsi* larvae in host nests with eggs rather than hatchlings occurred at the end of a four-year drought on the Galápagos in 2006. Since 2012, up to 80% of host nests with eggs may contain *P. downsi* larvae on Santa Cruz Island. Larval feeding by *P. downsi* on adult birds has been observed in laboratory finches and in one *Philornis* system (species unknown) in Puerto Rico. In light of changes in *P. downsi* larval feeding behaviour, we provided a description and photos of the larval instars for use in field identification. We compiled the observations to date of *Philornis* behaviour and ontogeny within a broad taxonomic framework and summarised patterns of change in the oviposition behaviour of *P. downsi* in its (presumably) novel habitat on the Galápagos Islands. By examining *P. downsi* in relation to other *Philornis* species, we provided a broad phylogenetic context for the potential behavioural repertoire of an invasive species under conditions of intense natural selection in a novel environment.

Acknowledgements

We thank the Galápagos National Park authority for research permits and the opportunity to work on the Galápagos, and the Charles Darwin Research Station for logistical support. We thank Charlotte Causton, Paola Lahuatte, Birgit Fessl, George Heimpel and Arno Cimadom for their useful comments on the manuscript. We thank Bradley Sinclair for advice on larval instar morphology. We thank Justin Holder, Grant Gully and Ben Parslow for their assistance with the photographs and guidance on using the Visionary System. This publication is contribution number 2277 of the Charles Darwin Foundation for the Galápagos Islands.

Author details

Lauren K. Common[1], Rachael Y. Dudaniec[2], Diane Colombelli-Négrel[1] and Sonia Kleindorfer[1,3]*

1 College of Science and Engineering, Flinders University, Adelaide, Australia

2 Department of Biological Sciences, Macquarie University, Sydney, Australia

3 Konrad Lorenz Research Center and Department of Behavioural Biology, University of Vienna, Austria

*Address all correspondence to: sonia.kleindorfer@flinders.edu.au

References

[1] Couri MS. Myiasis caused by obligatory parasites. 5a. Philornis Meinert (Muscidae). Myiasis in man and animals in the Neotropical region. In: Editora Pleiade. Brazil: Sao Paulo; 1999. pp. 44-70

[2] Teixeira DM. Myiasis caused by obligatory parasites. 5b. General observations on the biology of species of the genus Philornis Meinert, 1890 (Diptera, Muscidae). Myiasis in man and animals in the Neotropical Region. In: Editora Pleiade. Brazil: Sao Paulo; 1999. pp. 71-96

[3] Sabrosky CW, Bennett GF, Whitworth TL. Bird Blow Flies (*Protocalliphora*) in North America (Diptera: Calliphoridae) with Notes on Palearctic Species. Smithsonian Institution Press; 1989

[4] Pont AC. Revision of the genus *Passeromyia* Rodhain & Villeneuve (Diptera: Muscidae). In: Bulletin of the British Museum (Natural History) Entomology. 1974

[5] Bulgarella M, Heimpel GE. Host range and community structure of avian nest parasites in the genus *Philornis* (Diptera: Muscidae) on the island of Trinidad. Ecology and Evolution. 2015;**5**(17):3695-3703. DOI: 10.1002/ece3.1621

[6] McNew SM, Clayton DH. Alien invasion: Biology of *Philornis* flies highlighting *Philornis downsi*, an introduced parasite of Galápagos birds. Annual Review of Entomology. 2018;**63**:369-387. DOI: 10.1146/annurev-ento-020117-043103

[7] Dudaniec RY, Kleindorfer S. Effects of the parasitic flies of the genus *Philornis* (Diptera: Muscidae) on birds. Emu-Austral Ornithology. 2006;**106**(1):13-20. DOI: 10.1071/MU04040

[8] Bulgarella M, Quiroga MA, Heimpel GE. Additive negative effects of *Philornis* nest parasitism on small and declining Neotropical bird populations. Bird Conservation International. 2018;**29**(3):1-22. DOI: 10.1017/S0959270918000291

[9] Skidmore P. The biology of the Muscidae of the world (Series Entomologica). 1st ed. 550 pp. Dordrecht: Dr W; 1985

[10] Couri MS, Carvalho CD. Systematic relations among *Philornis* Meinert, *Passeromyia* Rodhain & Villeneuve and allied genera (Diptera, Muscidae). Brazilian Journal of Biology. 2003;**63**(2):223-232. DOI: 10.1590/S1519-69842003000200007

[11] Kutty SN, Pont AC, Meier R, Pape T. Complete tribal sampling reveals basal split in Muscidae (Diptera), confirms saprophagy as ancestral feeding mode, and reveals an evolutionary correlation between instar numbers and carnivory. Molecular Phylogenetics and Evolution. 2014;**78**:349-364. DOI: 10.1016/j.ympev.2014.05.027

[12] Haseyama KL, Wiegmann BM, Almeida EA, de Carvalho CJ. Say goodbye to tribes in the new house fly classification: A new molecular phylogenetic analysis and an updated biogeographical narrative for the Muscidae (Diptera). Molecular Phylogenetics and Evolution. 2015;**89**:1-2. DOI: 10.1016/j.ympev.2015.04.006

[13] Edworthy AB, Langmore NE, Heinsohn R. Native fly parasites are the principal cause of nestling mortality in endangered Tasmanian pardalotes. Animal Conservation. 2010;**22**(1):96-103

[14] Fessl B, Couri MS, Tebbich S. *Philornis downsi* Dodge & Aitken, new to the

Galápagos Islands (Diptera, Muscidae). Studia Dipterologica. 2001;**8**(1):317-322

[15] Fessl B, Heimpel GE, Causton CE. Invasion of an avian nest parasite, *Philornis downsi*, to the Galápagos Islands: Colonization history, adaptations to novel ecosystems, and conservation challenges. In: Disease Ecology. Cham: Springer; 2018. pp. 213-266. DOI: 10.1007/978-3-319-65909-1_9

[16] Kleindorfer S, Dudaniec RY. Host-parasite ecology, behavior and genetics: A review of the introduced fly parasite *Philornis downsi* and its Darwin's finch hosts. BMC Zoology. 2016;**1**(1):1. DOI: 10.1186/s40850-016-0003-9

[17] Lethier H, Bueno P. Report on the Reactive Monitoring Mission to Galápagos Islands World Heritage Site (Ecuador). IUCN. 2018. Available from: https://whc.unesco.org/en/documents/167914/ [Accessed: 29 June 2019]

[18] Toral-Granda MV, Causton CE, Jäger H, Trueman M, Izurieta JC, Araujo E, et al. Alien species pathways to the Galapagos Islands, Ecuador. PLoS One. 2017;**12**(9):e0184379. DOI: 10.1371/journal.pone.0184379

[19] Watkins G, Cruz F. Helmsley Charitable Trust's Galapagos Strategic Plan 2012.

[20] Causton CE, Cunninghame F, Tapia W. Management of the avian parasite *Philornis downsi* in the Galápagos Islands: A collaborative and strategic action plan. Galápagos Report. 2012;**2013**:167-173

[21] Fessl B, Tebbich S. *Philornis downsi*—A recently discovered parasite on the Galápagos archipelago—A threat for Darwin's finches? Ibis. 2002;**144**(3):445-451. DOI: 10.1046/j.1474-919X.2002.00076.x

[22] Causton CE, Peck SB, Sinclair BJ, Roque-Albelo L, Hodgson CJ, Landry B. Alien insects: Threats and implications for conservation of Galápagos Islands. Annals of the Entomological Society of America. 2006;**99**(1):121-143. DOI: 10.1603/0013-8746(2006)099[0121:AITAIF]2.0.CO;2

[23] Kleindorfer S, Sulloway FJ. Naris deformation in Darwin's finches: Experimental and historical evidence for a post-1960s arrival of the parasite *Philornis downsi*. Global Ecology and Conservation. 2016;**7**:122-131. DOI: 10.1016/j.gecco.2016.05.006

[24] Macquart J. Notice sur une nouvelle espèce d Aricie, diptère de la tribu des Anthomyzides. Annales de la Société Entomologique de France. 1854;**3**(1):657-660

[25] Meinert F. *Philornis molesta*, en paa Fugle snyltend Tachinarie. Videnskabelige Meddelelser fra den Naturhistoriske Forening I Kjøbenhavn. 1890;**1**(5):304-317

[26] Becker T, Bezzi M, Kertész K, Stein P. Katalog der Paläarktischen Dipteran. Cyclorrapha Aschiza. Acyclorrapha Schizophora: Schizommetopa. Budapest: Hódmezövåsárhely, Wesselényi; 1907;**3**:1-597. Availabe from: https://archive.org/details/katalogderpala03beck/page/n5

[27] Malloch JR. Notes on some of van der Wulp's species of north American Anthomyiidae (Diptera). Entomological News. 1921;**32**:40-45

[28] Aldrich JM. The genus *Philornis*-a bird-infesting group of Anthomyiidae. Annals of the Entomological Society of America. 1923;**16**(4):304-309. DOI: 10.1093/aesa/16.4.304

[29] Dodge HR. A new *Philornis* with coprophagous larva, and some related species (Diptera: Muscidae). Journal of the Kansas Entomological Society. 1963:239-247

[30] Dodge HR, Aitken TH. *Philornis* flies from Trinidad (Diptera: Muscidae). Journal of the Kansas Entomological Society. 1968;**41**(1):134-154

[31] Couri MS. Notes and descriptions of *Philornis* flies (Diptera, Muscidae, Cyrtoneurininae). Revista Brasileira de Entomologia. 1984;**43**(3):297-310

[32] Savage J, Vockeroth JR. Muscidae (house flies, stable flies). In: Brown BV, Borkent A, Cumming JM, Wood DM, Woodley NE, Zumbado MA, editors. Manual of Central American Diptera. Vol. 2. NRC Research Press; 2010. pp. 1281-1295

[33] Couri MS, De Carvalho CJB, Löwenberg-Neto P. Phylogeny of *Philornis* Meinert species (Diptera: Muscidae). Zootaxa. 2007;**1530**:19-26

[34] Couri MS, Antoniazzi LR, Beldomenico P, Quiroga M. Argentine *Philornis* Meinert species (Diptera: Muscidae) with synonymic notes. Zootaxa. 2009;**2261**(5262):77132

[35] Quiroga MA, Monje LD, Arrabal JP, Beldomenico PM. New molecular data on subcutaneous *Philornis* (Diptera: Muscidae) from southern South America suggests the existence of a species complex. Revista Mexicana de Biodiversidad. 2016;**87**(4):1383-1386. DOI: 10.1016/j.rmb.2016.10.018

[36] Teixeira DM, Couri MS, Luigi G. Notes on the biology of *Philornis rufoscutellaris* Couri, 1983 (Diptera, Muscidae) and on its association with nestling birds. Revista Brasileira de Entomologia. 1990;**34**(2):271-275

[37] Couri MS, Murphy TG, Hoebeke R. *Philornis fasciventris* (Wulp) (Diptera: Muscidae): Description of the male, larva and puparium, with notes on biology and host association. Neotropical Entomology. 2007;**36**(6):889-893. DOI: 10.1590/S1519-566X2007000600009

[38] Leite GA, Matsui QY, Couri MS, Monteiro AR. New association between *Philornis* Meinert (Diptera: Muscidae) and Falconidae (Aves: Falconiformes). Neotropical Entomology. 2009;**38**(5):686-687. DOI: 10.1590/S1519-566X2009000500021

[39] Bulgarella M, Quiroga MA, Boulton RA, Ramírez IE, Moon RD, Causton CE, et al. Life cycle and host specificity of the parasitoid *Conura annulifera* (hymenoptera: Chalcididae), a potential biological control agent of *Philornis downsi* (Diptera: Muscidae) in the Galápagos Islands. Annals of the Entomological Society of America. 2017;**110**(3):317-328. DOI: 10.1093/aesa/saw102

[40] Proudfoot GA, Teel PD, Mohr RM. Ferruginous pygmy-owl (*Glaucidium brasilianum*) and eastern screech-owl (*Megascopes asio*): New hosts for *Philornis mimicola* (Diptera: Muscidae) and *Ornithodoros concanensis* (Acari: Argasidae). Journal of Wildlife Diseases. 2006;**42**(4):873-876. DOI: 10.7589/0090-3558-42.4.873

[41] Spalding MG, Mertins JW, Walsh PB, Morin KC, Dunmore DE, Forrester DJ. Burrowing fly larvae (*Philornis porteri*) associated with mortality of eastern bluebirds in Florida. Journal of Wildlife Diseases. 2002;**38**(4):776-783. DOI: 10.7589/0090-3558-38.4.776

[42] Le Gros A, Stracey CM, Robinson SK. Associations between northern mockingbirds and the parasite *Philornis porteri* in relation to urbanization. The Wilson Journal of Ornithology. 2011;**123**(4):788-796. DOI: 10.1676/10-049.1

[43] Kinsella JM, Winegarner CE. Notes on the life history of *Neomusca porteri* (Dodge), parasitic on nestlings of the great crested flycatcher in Florida. Journal of Medical Entomology. 1974;**11**(5):633

[44] Fessl B, Sinclair BJ, Kleindorfer S. The life-cycle of *Philornis downsi* (Diptera: Muscidae) parasitizing Darwin's finches and its impacts on nestling survival. Parasitology. 2006;**133**(6):739-747. DOI: 10.1017/S0031182006001089

[45] O'Connor JA, Robertson J, Kleindorfer S. Video analysis of host–parasite interactions in nests of Darwin's finches. Oryx. 2010;**44**(4):588-594. DOI: 10.1017/S0030605310000086

[46] O'Connor JA, Robertson J, Kleindorfer S. Darwin's finch begging intensity does not honestly signal need in parasitised nests. Ethology. 2014;**120**(3):228-237. DOI: /10.1111/eth.12196

[47] Young BE. Effects of the parasitic botfly *Philornis carinatus* on nestling house wrens, *Troglodytes aedon*, in Costa Rica. Oecologia. 1993;**93**(2):256-262. DOI: 10.1007/BF00317679

[48] Arendt WJ. *Philornis* ectoparasitism of pearly-eyed thrashers. I. Impact on growth and development of nestlings. The Auk. 1985;**102**(2):270-280. DOI: 10.2307/4086769

[49] Arendt WJ. *Philornis* ectoparasitism of pearly-eyed thrashers. II. Effects on adults and reproduction. The Auk. 1985;**102**(2):281-292. DOI: 10.2307/4086770

[50] Rabuffetti FL, Reboreda JC. Early infestation by bot flies (*Philornis seguyi*) decreases chick survival and nesting success in chalk-browed mockingbirds (*Mimus saturninus*). The Auk. 2007;**124**(3):898-906. DOI: 10.1642/0004-8038(2007)124[898:EIBBFP]2.0.CO;2

[51] Quiroga MA, Reboreda JC. Lethal and sublethal effects of botfly (*Philornis seguyi*) parasitism on house wren nestlings. The Condor. 2012;**114**(1): 197-202. DOI: /10.1525/cond.2012.110152

[52] Silvestri L, Antoniazzi LR, Couri MS, Monje LD, Beldomenico PM. First record of the avian ectoparasite *Philornis downsi* Dodge & Aitken, 1968 (Diptera: Muscidae) in Argentina. Systematic Parasitology. 2011;**80**(2):137. DOI: 10.1007/s11230-011-9314-y

[53] Cimadom A, Ulloa A, Meidl P, Zöttl M, Zöttl E, Fessl B, et al. Invasive parasites, habitat change and heavy rainfall reduce breeding success in Darwin's finches. PLoS One. 2014;**9**(9):e107518. DOI: 10.1371/journal.pone.0107518

[54] Kleindorfer S, Peters KJ, Custance G, Dudaniec RY, O'Connor JA. Changes in *Philornis* infestation behavior threaten Darwin's finch survival. Current Zoology. 2014;**60**(4):542-550. DOI: 10.1093/czoolo/60.4.542

[55] Uhazy LS, Arendt WJ. Pathogenesis associated with philornid myiasis (Diptera: Muscidae) on nestling pearly-eyed thrashers (Aves: Mimidae) in the Luquillo Rain Forest, Puerto Rico. Journal of Wildlife Diseases. 1986;**22**(2):224-237. DOI: 10.7589/0090-3558-22.2.224

[56] Lincango P, Causton C. Crianza en cautiverio de *Philornis downsi*, en las Islas Galápagos. Charles Darwin Fourndation: Informe interno; 2008

[57] Lincango P, Causton C, Cedeño D, Castañeda J, Hillstrom A, Freund D. Interactions between the avian parasite, *Philornis downsi* (Diptera: Muscidae) and the Galápagos flycatcher, *Myiarchus magnirostris* Gould (Passeriformes: Tyrannidae). Journal of Wildlife Diseases. 2015;**51**(4):907-910. DOI: 10.7589/2015-01-025

[58] Kleindorfer S, Peters KJ, Hohl L, Sulloway FJ. Flight behaviour of an introduced parasite affects its Galápagos Island hosts: *Philornis downsi* and Darwin's finches. Chapter 10 In:

Weis JS, Sol D, editors. Biological Invasions and Animal Behaviour. Cambridge: Cambridge University Press; 2016

[59] Cha DH, Mieles AE, Lahuatte PF, Cahuana A, Lincango MP, Causton CE, et al. Identification and optimization of microbial attractants for Philornis downsi, an invasive fly parasitic on Galápagos birds. Journal of Chemical Ecology. 2016;**42**(11):1101-1111. DOI: 10.1007/s10886-016-0780-1

[60] Bateman AJ. Intra-sexual selection in *Drosophila*. Heredity. 1948;**2**(3):349-368

[61] Fowler K, Partridge L. A cost of mating in female fruitflies. Nature. 1989;**338**(6218):760. DOI: 10.1038/338760a0

[62] O'Connor JA, Sulloway FJ, Robertson J, Kleindorfer S. *Philornis downsi* parasitism is the primary cause of nestling mortality in the critically endangered Darwin's medium tree finch (*Camarhynchus pauper*). Biodiversity and Conservation. 2010;**19**(3):853-866. DOI: 10.1007/s10531-009-9740-1

[63] Kleindorfer S, O'Connor JA, Dudaniec RY, Myers SA, Robertson J, Sulloway FJ. Species collapse via hybridization in Darwin's tree finches. The American Naturalist. 2014;**183**(3):325-341. DOI: 10.1086/674899

[64] Ramirez I. *Philornis downsi* interactions with its host in the introduced range and its parasitoids in its native range [thesis]. The University of Minnesota; 2018

[65] Dudaniec RY, Gardner MG, Kleindorfer S. Offspring genetic structure reveals mating and nest infestation behaviour of an invasive parasitic fly (*Philornis downsi*) of Galápagos birds. Biological Invasions. 2010;**12**(3):581-592. DOI: 10.1007/s10530-009-9464-x

[66] Carlson DA, Mayer MS, Silhacek DL, James JD, Beroza M, Bierl BA. Sex attractant pheromone of the house fly: Isolation, identification and synthesis. Science. 1971;**174**(4004):76-78. DOI: 10.1126/science.174.4004.76

[67] Chapman JW, Knapp JJ, Howse PE, Goulson D. An evaluation of (Z)-9-tricosene and food odours for attracting house flies, *Musca domestica,* to baited targets in deep-pit poultry units. Entomologia Experimentalis et Applicata. 1998;**89**(2):183-192. DOI: 10.1046/j.1570-7458.1998.00398.x

[68] Jiang Y, Lei CL, Niu CY, Fang YL, Xiao C, Zhang ZN. Semiochemicals from ovaries of gravid females attract ovipositing female houseflies, *Musca domestica*. Journal of Insect Physiology. 2002;**48**(10):945-950. DOI: 10.1016/S0022-1910(02)00162-2

[69] Collignon RM. Semiochemicals of *Philornis downsi* (Dipter: Muscidae), a Parasite of Passerine Birds of the Galápagos Islands. State University of New York College of Environmental Science and Forestry; 2011

[70] Doherty KM. Chemical attractants of *Philornis downsi* (Diptera: Muscidae), an invasive parasite of birds in the Galápagos Islands [honors thesis]. SUNY College of Environmental Science and Forestry; 2012

[71] Collignon R, Boroczky K, Mieles AE, Causton CE, Lincango MP, Teale SA. Cuticular lipids and mate attraction in the avian parasite *Philornis downsi* (Diptera: Muscidae). In: International Society of Chemical Ecology; Poster; 8-12 July 2014. University of Illinois at Urbana-Champaign

[72] Lance DR, McInnis DO. Biological basis of the sterile insect technique. In: Sterile Insect Technique. Dordrecht:

Springer; 2005. pp. 69-94. DOI: 10.1007/1-4020-4051-2_3

[73] Meier R, Kotrba M, Ferrar P. Ovoviviparity and viviparity in the Diptera. Biological Reviews. 1999;**74**(3):199-258

[74] Patitucci LD, Quiroga M, Couri MS, Saravia-Pietropaolo MJ. Oviposition in the bird parasitic fly *Philornis torquans* (Nielsen, 1913) (Diptera: Muscidae) and eggs' adaptations to dry environments. Zoologischer Anzeiger. 2017;**267**:15-20. DOI: 10.1016/j.jcz.2017.01.004

[75] Lahuatte PF, Lincango MP, Heimpel GE, Causton CE. Rearing larvae of the avian nest parasite, *Philornis downsi* (Diptera: Muscidae), on chicken blood-based diets. Journal of Insect Science. 2016;**16**(1):84. DOI: 10.1093/jisesa/iew064

[76] Couri MS, Barbosa L, Marini MÂ, Duca C, Pujol-Luz JR. A new host for *Philornis torquans* (Diptera, Muscidae) from the Brazilian Cerrado. Papéis Avulsos de Zoologia. 2018;**58**:e20185857. DOI: 10.11606/1807-0205/2018.58.57

[77] Ben-Yosef M, Zaada DS, Dudaniec RY, Pasternak Z, Jurkevitch E, Smith RJ, et al. Host-specific associations affect the microbiome of *Philornis downsi*, an introduced parasite to the Galápagos Islands. Molecular Ecology. 2017;**26**(18):4644-4656. DOI: 10.1111/mec.14219

[78] Segura LN, Reboreda JC. Botfly parasitism effects on nestling growth and mortality of red-crested cardinals. The Wilson Journal of Ornithology. 2011;**123**(1):107-115. DOI: 10.1676/10-053.1

[79] Couri MS, Rabuffetti FL, Reboreda JC. New data on *Philornis seguyi* Garcia (1952) (Diptera, Muscidae). Brazilian Journal of Biology. 2005;**65**(4):631-637. DOI: 10.1590/S1519-69842005000400010

[80] Nores AI. Botfly ectoparasitism of the Brown Cacholote and the firewood-gatherer. The Wilson Bulletin. 1995;**107**(4):734-738

[81] De la Peña MR, Beldoménico PM, Antoniazzi L. Pichones de aves parasitados por larvas de *Philornis* sp. (Diptera: Muscidae) en un sector de la provincia biogeográfica del Espinal de Santa Fe, Argentina. Revista FAVE–Ciencias Veterinarias. 2003;**2**:141-146

[82] Oniki Y. Notes on fly (Muscidae) parasitism of nestlings of south American birds. Gerfaut. 1983;**73**:281-286

[83] Mendonça ED, Couri MS. New associations between *Philornis* Meinert (Diptera, Muscidae) and Thamnophilidae (Aves, Passeriformes). Revista Brasileira de Zoologia. 1999;**16**(4):1223-1225. DOI: 10.1590/S0101-81751999000400030

[84] Herrera JM, Bermúdez SE. Miasis ocasionada por *Philornis* spp. (Diptera: Muscidae) in *Dendroica castanea* (Aves: Parulidae) en Panamá. Revista mexicana de biodiversidad. 2012;**83**(3):854-855. DOI: 10.7550/rmb.25650

[85] Cimadom A, Causton C, Cha DH, Damiens D, Fessl B, Hood-Nowotny R, et al. Darwin's finches treat their feathers with a natural repellent. Scientific Reports. 2016;**6**:34559. DOI: 10.1038/srep34559

[86] Cimadom A, Jäger H, Schulze CH, Hood-Nowotny R, Wappl C, Tebbich S. Weed management increases the detrimental effect of an invasive parasite on arboreal Darwin's finches. Biological Conservation. 2019;**233**:93-101. DOI: 10.1016/j.biocon.2019.02.025

[87] Huber SK, Owen JP, Koop JA, King MO, Grant PR, Grant BR, et al. Ecoimmunity in Darwin's finches: Invasive

parasites trigger acquired immunity in the Medium Ground Finch (*Geospiza fortis*). PLoS One. 2010;**5**(1):e8605. DOI: 10.1371/journal.pone.0008605

[88] Koop JA, Owen JP, Knutie SA, Aguilar MA, Clayton DH. Experimental demonstration of a parasite-induced immune response in wild birds: Darwin's finches and introduced nest flies. Ecology and Evolution. 2013;**3**(8):2514-2523. DOI: 10.1002/ece3.651

[89] Sage R, Boulton RA, Lahuatte PF, Causton CE, Cloutier R, Heimpel GE. Environmentally cued hatching in the bird-parasitic nest fly *Philornis downsi*. Entomologia Experimentalis et Applicata. 2018;**166**(9):752-760. DOI: 10.1111/eea.12721

[90] Dudaniec RY, Kleindorfer S, Fessl B. Effects of the introduced ectoparasite *Philornis downsi* on haemoglobin level and nestling survival in Darwin's small ground finch (Geospiza fuliginosa). Austral Ecology. 2006;**31**(1):88-94. DOI: 10.1111/j.1442-9993.2006.01553.x

[91] Fessl B, Kleindorfer S, Tebbich S. An experimental study on the effects of an introduced parasite in Darwin's finches. Biological Conservation. 2006;**127**(1):55-61. DOI: 10.1016/j.biocon.2005.07.013

[92] Dudaniec RY, Fessl B, Kleindorfer S. Interannual and interspecific variation in intensity of the parasitic fly, *Philornis downsi*, in Darwin's finches. Biological Conservation. 2007;**139**(3-4):325-332. DOI: 10.1016/j.biocon.2007.07.006

[93] Wiedenfeld DA, Fessl B, Kleindorfer S, Valarezo JC. Distribution of the introduced parasitic fly Philornis downsi (Diptera, Muscidae) in the Galápagos Islands. Pacific Conservation Biology. 2007;**13**(1):14-19. DOI: 10.1071/PC070014

[94] Kleindorfer S, Dudaniec RY. Love thy neighbour? Social nesting pattern, host mass and nest size affect ectoparasite intensity in Darwin's tree finches. Behavioral Ecology and Sociobiology. 2009;**63**(5):731-739. DOI: 10.1007/s00265-008-0706-1

[95] O'Connor JA, Dudaniec RY, Kleindorfer S. Parasite infestation and predation in Darwin's small ground finch: Contrasting two elevational habitats between islands. Journal of Tropical Ecology. 2010;**26**(3):285-292. DOI: 10.1017/S0266467409990678

[96] Koop JA, Huber SK, Laverty SM, Clayton DH. Experimental demonstration of the fitness consequences of an introduced parasite of Darwin's finches. PLoS One. 2011;**6**(5):e19706. DOI: 10.1371/journal.pone.0019706

[97] Koop JA, Le Bohec C, Clayton DH. Dry year does not reduce invasive parasitic fly prevalence or abundance in Darwin's finch nests. Reports in Parasitology. 2013;**3**:11-17. DOI: 10.2147/RIP.S48435

[98] Knutie SA, Koop JA, French SS, Clayton DH. Experimental test of the effect of introduced hematophagous flies on corticosterone levels of breeding Darwin's finches. General and Comparative Endocrinology. 2013;**193**:68-71. DOI: 10.1016/j.ygcen.2013.07.009

[99] Knutie SA, McNew SM, Bartlow AW, Vargas DA, Clayton DH. Darwin's finches combat introduced nest parasites with fumigated cotton. Current Biology. 2014;**24**(9):R355-R356. DOI: 10.1016/j.cub.2014.03.058

[100] Knutie SA, Owen JP, McNew SM, Bartlow AW, Arriero E, Herman JM, et al. Galápagos mockingbirds tolerate introduced parasites that affect Darwin's finches. Ecology. 2016;**97**(4):940-950. DOI: 10.1890/15-0119

[101] Heimpel GE, Hillstrom A, Freund D, Knutie SA, Clayton DH. Invasive parasites and the fate of Darwin's finches in the Galápagos Islands: The case of the vegetarian finch (*Platyspiza crassirostris*). The Wilson Journal of Ornithology. 2017;**129**(2):345-349. DOI: 10.1676/16-050.1

[102] Knutie SA. Relationships among introduced parasites, host defenses, and gut microbiota of Galapagos birds. Ecosphere. 2018;**9**(5):e02286. DOI: 10.1002/ecs2.2286

[103] McNew SM, Knutie SA, Goodman GB, Theodosopoulos A, Saulsberry A, Yépez RJ, et al. Annual environmental variation influences host tolerance to parasites. Proceedings of the Royal Society B. 2019;**286**(1897):20190049. DOI: 10.1098/rspb.2019.0049

[104] Peters KJ, Evans C, Aguirre JD, Kleindorfer S. Genetic admixture predicts parasite intensity: Evidence for increased hybrid performance in Darwin's tree finches. Royal Society Open Science. 2019;**6**(4):181616. DOI: 10.1098/rsos.18161

Chapter 3

Life Cycle and Cytogenetic Study of Mosquitoes (Diptera: Culicidae)

Sahar Abd

Abstract

Mosquitoes fall into the Culicidae family of the order Diptera within class Insecta and members of the phylum Arthropod. This family includes two important medical and veterinary *important* disease vectors due to their roles for transmission of various viruses, bacteria, and parasites—Anophelinae and Culicinae. The mosquitoes undergo four stages of transformation during their lifetime: egg, larva, pupa, and adult. These have complete metamorphoses or so called Holometabola. Commonly known as the southern house mosquito, *Culex quinquefasciatus* Say is a medium-sized brown insect that exists throughout the tropics and the lower latitudes of temperate regions, and a vector of many pathogens of humans as well as both domestic and wild animals. Although an intensified interest in mosquito cytogenetics in the past decade has produced a number of contributions to knowledge on this subject, the available information is still superficial and limited to a few mosquito species only. Therefore, the karyotype of the populations of the mosquito *C. quinquefasciatus* has been studied collected from three provinces: Babylon, Baghdad, and Wasit of Iraq. The study showed that the chromosomes karyotyping of this species consisted of three pairs of chromosomes (i.e., 2n = 6). In conclusion, it is stressed that prospects are especially good for evolutionary and genetic studies involving chromosomal polymorphism.

Keywords: mosquitoes, Diptera, Culex, *Aedes* and *Anopheles*

1. Introduction

Female mosquitoes attack humans and animals to obtain the blood necessary to sustain their vital activities and make eggs. They use all warm-blooded animals, causing severe uncomfortable and serious harm due to loss of blood, itching, and allergies as well as transmission of pathogens [1].

Carbon dioxide, which is emitted from respiration as well as sweat, is an attractive substance for mosquitoes [2]. The symptoms or reactions caused by mosquito bites are pink rash around the attacked spot accompanied by itching and pain, and may be accompanied by symptoms of allergies. The saliva of mosquitoes contains chemical components which cause these symptoms on the hosts. *Staphylococcus aureus* found in the salivary glands secretes antigenic substances, agglutinins and anticoagulants, and it has been found that the saliva of mosquitoes prevents the growth of chicken embryos and causes death if infected. Feeding by the mosquito is through the parts of her mouth reaching the capillary blood vessels, which leads to puncture in several areas, causing slow or stop of the blood flow to produce perfusion and bleeding appears in the form of small bleeding spots, especially in the areas around the eye in children [3].

Mosquitoes attack exposed parts of the nose, ears, and limbs, and there is record of a few hundred *C. quinquefasciatus* mosquitoes per person in 1 h, as *C. quinquefasciatus* prefers blood absorption from the face and trunk [4]. Also found among the areas of the body preferred by mosquitoes for nutrition is the foot more than the leg because the region is characterized by smells resulting from the frequent presence of sweat glands [5].

The nature of mosquito adults in their frequency to a number of families for feeding makes these important to transport pathogens from the infected person or animal to a healthy person or animal [6]. The mosquitoes occupy an important place among the medical insects carrying dangerous pathogens that lead to human death. The most serious of these causes are the cases of malaria, which is due to the species of *Plasmodium* that is transmitted by the female sex species. The *Anopheles* species show a vital transmission of malaria in the world and 380 species of *Anopheles* mosquito have been known as vectors. For malaria, about 60 mosquito species are attracted to humans for nutrition [7].

At least 20 species of malaria pathogens have been registered in South Asia, including India, Pakistan, Bangladesh, and Sri Lanka. The most serious pathogens transmitting species are *Anopheles stephensi* Liston, *Anopheles sacharovi* Favre, and *Anopheles gambiae* Giles [8]. Filariasis lymphatic disease is a serious disease that causes three types of worms: *Wuchereria bancrofti, Brugia malayi*, and *B. timori* are transmitted by genus *Culex*, *Aedes*, and *Anopheles*. But the most dangerous of these types of worms is *W. bancrofti* [9].

The *Culex* mosquitoes also transmit many pathogens that cause serious diseases to humans, especially those that cause meningitis in humans and animals, and from these initiating West Nile virus and meningitis, *Louis encephalitis virus, Japanese encephalitis virus,* and *chicken pox* [10, 11].

Yellow fever, which is a short-term acute illness, often causes death and starts with fever, headaches, and jaundice. The patient's color becomes yellow, internal bleeding and vomiting can occur, and death may take place within 3 days. This type of disease is transmitted by *Culex* and *Aedes* mosquitoes. Other serious diseases whose pathogens are mosquitoes include dengue fever and dengue hemorrhagic fever. The *Aedes aegypti* (*Linnaeus*), the yellow fever mosquito, and *Aedes albopictus* (Skuse), the Asian tiger mosquito, are the main vectors of these diseases in tropical and subtropical regions. A. *aegypti* is the major carrier of the disease in East Asian and warmer countries [12].

2. Classification of mosquito

Kingdom: Animalia
Phylum: Arthropoda
Subphylum: Hexapoda
Class: Insecta
Subclass: Pterygota
Order: Diptera
Suborder: Nematocera
Family: Culicidae
Subfamily: Culicinae [13]

The fragmentary data available on mosquito life cycle and cytogenetics clearly indicate the kind and amount of future work to be done. Out of so many kinds of mosquitos thus far described, there have been life cycle and cytogenetics information on fewer species and much of this is very superficial. Therefore, the life cycle and the karyotype of the populations of the mosquito *C. quinquefasciatus* have been studied collected from three provinces of Iraq.

3. Life cycle of mosquito

Mosquitoes fall in the Culicidae family of the Diptera order. This family includes two medically important subfamilies: Anophelinae and Culicinae [14]. The Culicidae family has about 3500 species of mosquitoes belonging to 43 genera, including *Culex* [15]. *Culex* is a cosmopolitan genus and one of the largest groups of the family Culicidae with 768 species divided among 26 subgenera [16]. Classification under the *Culex* is based on the diagnostic characteristics of females and males mosquito.

However, most of the available classification keys should be used with caution because the phenotypic traits of females may be either multiform or overlapping. The morphological characteristics of the fourth larval stage are also used to determine species; however, some overlaps between species may also be difficulties for the researchers to obtain an accurate classification of mosquito species. Despite the significant phenotypic similarity between the *northern house mosquito* or *common house mosquito Culex pipiens* Linnaeus complex, they differ widely in behavior, physiology, and preference.

Mosquitoes are widespread in all the tropical and subtropical regions of the world, which extend into the Arctic Circle but are absent in Antarctica [17]. The eggs from temperate breeds have more strain than those found in warmer regions [18]. Mosquitoes are found at a height of 550 m and a depth of 1250 m below sea level [19]. The shallow-water marshes containing plants are a preferred environment for the growth and reproduction of mosquitoes. The most important species that prefer these environments are the types of *Culex*, especially *C. pipiens* and *C. salinarius* Coquillett [20].

Most mosquitoes mate shortly after leaving the envelope of the pupa. The sperms are passed by the male and are saved by the female within the spermatheca and all eggs can be fertilized in the female throughout her life. Therefore, one mating is sufficient for each female throughout her lifetime [7].

Female mosquitoes need to obtain the blood meal necessary for egg growth and maturity. This is referred to as the development of anautogenous development, as in the case of type *C. quinquefasciatus*, but in some species where at least the first meal of eggs can develop. Probably, following the payments without a blood meal, the process is called autogenous development as in *Culex molestus* Forskal [15].

Female often need a blood meal either before or after mating before they mature. Many species draw on humans to get their meals from the blood and feed a little on human blood by preference for any other animal so these are called anthropophilic species in their dietary habits, while those that feed primarily on animals (mammals, reptiles, and birds) are animal lover zoophilic, and mosquitoes feeding on birds are called bird lovers ornithophilic [21].

Some mosquitoes prefer feeding at specific times of the day, such as dusk, dawn, or midnight, but daytime feeding can also occur. A few mosquitoes enter homes often for human feeding and are called endophagic feeders in their eating habits, while those that bite outside of their homes are called exophagic [7].

All mosquito species go through four stages during their life cycles (**Figure 1**):

- egg—hatches when exposed to water;

- larva—"wriggler" lives in water and most species come to surface to breathe air;
- pupa—"tumbler" does not feed and it is the stage just before emerging as an adult; and
- adult—flies short time after emerging and after its body parts have hardened.

The mosquitoes undergo four stages of transformation during their lifetime: egg, larva, pupa, and adult. These have complete metamorphoses or are called Holometabola (**Figure 1**). Females usually mate once but put eggs in several batches throughout their lives, the largest of which is the first batch. For a female to do so, she must feed more than once on the blood. The blood meal is digested 2–3 days in the tropics and more so in the temperate regions. Pregnant females look for suitable place to lay eggs. They feed on the blood to put another egg batch. This process is repeated until the mosquito dies, which may last up to 1 month. Eggs are usually brown and long cylindrical, placed vertically on the surface of the water and laid together to form an egg raft, which can contain 30–300 eggs (**Figure 2**). The *Aedes* females lay eggs individually on wet places and eggs hatch usually after 24–30 h. The larva (**Figure 3**) passes through four stages where it is hatched and molts three times, ranging from 1 to 3 days. The larva feeds in minutes amount

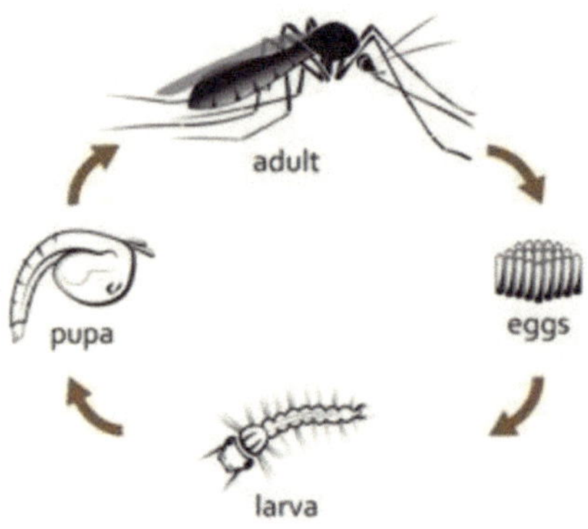

Figure 1.
Life cycle of mosquito.

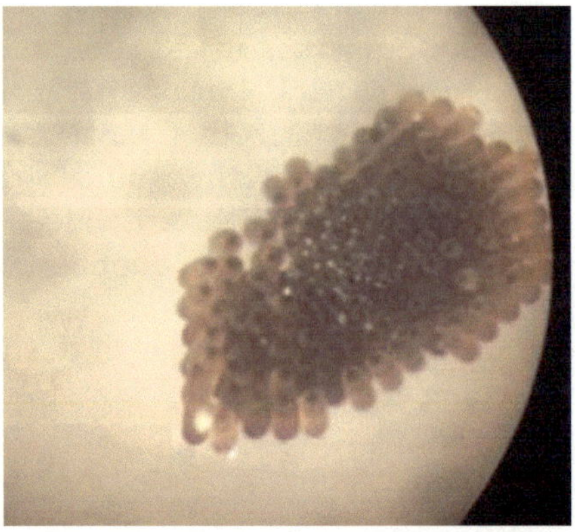

Figure 2.
Egg raft of Culex quinquefasciatus.

of organic matter as well as on bacteria, primates, lichens, fungi, and spores by means of a permanent circular motion around the mouth through mouth-like mouthwashes.

The larvae are transformed into the pupa and are of the type of active pupa (**Figure 4**), which differ in their appearance from the larvae. The large cephalothorax region includes the head and its tail represents the belly of the complete insect. The tail ends with a pair of paddles. Within water, the weight of the pupa is lighter than the weight of water, so they rise again to the surface of the water slowly through the raft or float. The pupa has a pair of trumpets respiratory connected directly to the open air and is used to breathe [11].

Figure 3.
Larva of Culex quinquefasciatus.

Figure 4.
Pupa of Culex quinquefasciatus.

The adult's length is between 3.96 and 4.25 mm (**Figure 5**), and it is are made up of a head with two large compound eyes, a thorax, a pair of scaled wings, and six jointed legs. The thorax, legs, and wings are covered with dark brown scales. The abdomen is covered with black scales and some of which are white, and the anal cercus is withdrawn. The adults rest on the surfaces in such a way that the thorax and head are parallel to the surface and the hose may form a small angle with the surface. The antenna of the male is plumose, easily visible near the eye and the antenna of the female is pilose [14].

Male mosquitoes last less than a week and the females may live for a couple of months and that is only with ideal conditions depending on how much warmth and moisture. Sometimes females have to wait for about a day for their reproductive parts to develop completely.

The role of mosquitoes life cycle to manage their populations is vital throughout the world and especially in the tropics because these spread many diseases. Generally, mosquitoes control operations are targeted against three different problems; nuisance mosquitoes bother people around homes, in parks, and recreational areas; economically important mosquitoes reduce real estate values consisting of land and the buildings, adversely affecting tourism and related business interests; and public health is the focus when mosquitoes are vectors or transmitters of infectious diseases [22–27]. Therefore, mosquitoes prevention and control involves a basic understanding of their life cycle, removal of potential egg laying sites, removal of any sources of standing water, keeping of weeds and other vegetation mowed and trimmed to minimize shelter for adult mosquitoes, considering stocking of mosquitoes eating fish for areas with ponds, control of adult mosquitoes, and personal protection [28–38].

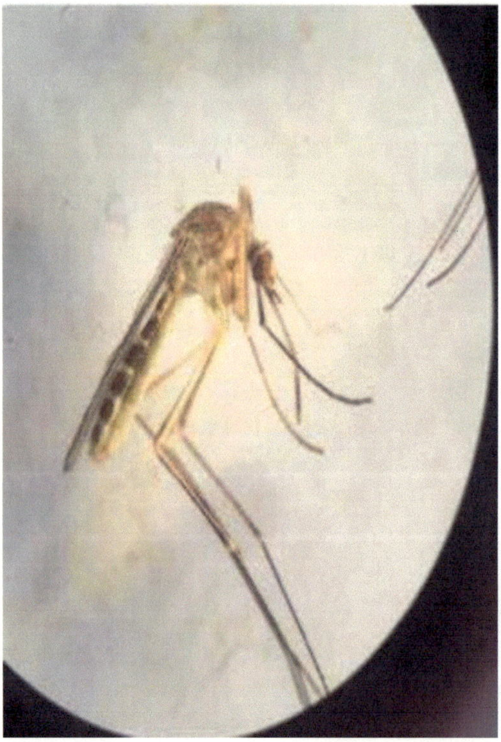

Figure 5.
Female adult of Culex quinquefasciatus.

4. Cytogenetic study

During the last so many years, there has been a renewed emphasis upon basic research in mosquito biology. The dramatic development of resistance over the world and the unequivocal demonstration of the genetic basis of this resistance have underlined the need for more research and increasingly intensive research on the genetics and cytogenetics of mosquitoes, and today, relatively little is known in this field.

The study shows that the chromosomes karyotyping of this species consists of three pairs of chromosomes (i.e., 2n = 6). There are one pair of sexual chromosomes and two pairs of autosomal chromosomes (**Figure 6**).

The average length for autosomal chromosomes collected from Baghdad province is 5.87 and 6.99 μM for chromosome 2 and chromosome 3, respectively, and the length of sexual chromosome is 4.46 μM, the average length for autosomal chromosomes collected from Babylon province is 5.51 and 6.96 μM for chromosome 2 to chromosome 3, respectively, and the length of sexual chromosome is 4.33 μM; while the average length for autosomal chromosomes collected from Wasit province is 5.5 and 6.97 μM for chromosome 2 and chromosome 3, respectively, and the length of sexual chromosome is 3.89 μM (**Table 1**).

When measuring the arm ratio for the autosomal and sexual chromosomes, it shows that the metacentric chromosomes show the average arm ratio for the chromosomes of the Baghdad province as 1.21, 1.28, and 1.25 μM for chromosomes 1, 2, and 3, respectively; the ratio of arm to autosomal and sexual chromosomes of Babylon as 1.18, 1.15, and 1.20 μM from chromosome 1 to chromosome 3, respectively; and the ratio of arm of the autosomal and sexual chromosomes of the Wasit province as 1.27, 1.20, and 1.30 μM from chromosome 1 to chromosome 3, respectively (**Table 2**).

In regard to the relative length for the autosomal chromosomes, the average length of the relative chromosomes of the Baghdad is 33.98 and 40.18% for chromosome 2 and chromosome 3, respectively, and the relative length of the sexual chromosome is 25.8%, while the relative length of the chromosomes of the Babylon is 32.32 and 41.42% in the case of chromosome 2 and chromosome 3, respectively; the relative length of the sexual chromosome is 25.77%. As for the Wasit province, the relative length of chromosomes is 33.61 and 42.6% in case of chromosome 2 and chromosome 3, respectively, and the rate of relative length of sexual chromosome is the 23.78% (**Table 3**).

When measuring the average ratio of centromere index for this insect of Baghdad province, it is 43.8 and 44.35 μM for chromosome 2 and chromosome 3, respectively, while the average ratio of the centromere index is 45.1 μM for the sexual

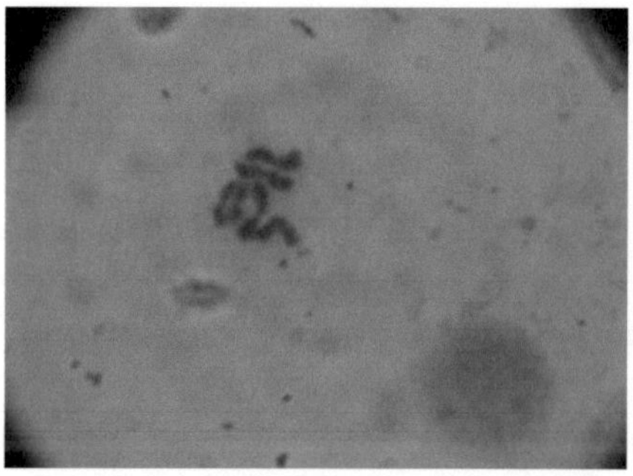

Figure 6.
Three pairs of chromosomes of Culex quinquefasciatus.

Region	Ch. No.		
	Ch. I* **M ± S.D**	**Ch. II**** **M ± S.D**	**Ch. III**** **M ± S.D**
Baghdad	4.46 ± 1 a	5.87 ± 1 a	6.99 ± 1 a
Babail	4.33 ± 1 a	5.51 ± 1.52 a	6.96 ± 1 a
Wasit	3.89 ± 1 a	5.5 ± 1.52 a	6.97 ± 1 a
	P = 0.7721	P = 0.7793	P = 1.0

*Sexual chromosomes.
**Mean autosomal chromosomes.

Table 1.
The average total length (μM) of chromosomes collected from different province in Iraq.

Region	Ch. No.		
	Ch. I* **M ± S.D**	**Ch. II**** **M ± S.D**	**Ch. III**** **M ± S.D**
Baghdad	1.21 ± 1 a	1.28 ± 1 a	1.25 ± 1 a
Babail	1.18 ± 1 a	1.1 a	1.20 ± 1 a
Wasit	1.27 ± 1 a	1.20 ± 1 a	1.30 ± 1 a
	p = 0.99	p = 0.99	p = 0.99

*Sexual chromosomes.
**Mean autosomal chromosomes.

Table 2.
The average arm ratio for the chromosomes of the different provinces in Iraq.

Region	Ch. No.		
	Ch. I* **M ± S.D**	**Ch. II**** **M ± S.D**	**Ch. III**** **M ± S.D**
Baghdad	25.8% ±10 a	33.98% ±6.08 a	40.18% ±8.54 a
Babail	25.77% ± 10 a	32.79% ±10 a	41.42% ±8.54 a
Wasit	23.78% ± 10 a	33.61% ±10 a	42.6% ±10 a
	p = 0.96	p = 0.83	p = 0.99

*Sexual chromosomes.
**Mean autosomal chromosomes.

Table 3.
The average length of the relative chromosomes of the Culex quinquefasciatus.

chromosome. The average ratio of the centromere index of the Babylon insect is 46.5 and 46.6 μM for chromosome 2 and chromosome 3, respectively, while the chromosome reached 45.72 μM for the sexual chromosome, and the average centromere index of the Wasit province is 45.1 and 43.32 μM for chromosome 2 and chromosome 3, respectively, and for sexual chromosome, it is 44 μM (**Tables 4** and **5**).

Region	Ch. No.		
	Ch. I* M ± S.D	Ch. II** M ± S.D	Ch. III** M ± S.D
Baghdad	45.1% ± 10 a	43.8% ± 10 a	44.35% ± 10 a
Babail	45.72% ± 10 a	46.5% ± 10 a	46.6% ± 10 a
Wasit	44% ± 10 a	45.1% ± 10 a	43.32% ± 10 a
	p = 0.924	p = 0.980	p = 0.951

**Sexual chromosomes.*
***Mean autosomal chromosomes.*

Table 4.
The average ratio of centromere index chromosomes of the Culex quinquefasciatus.

Region	Ch. No.		
	Ch. I*	Ch. II**	Ch. III**
Baghdad	M Metacentric	M Metacentric	M Metacentric
Babail	M Metacentric	M Metacentric	M Metacentric
Wasit	M Metacentric	M Metacentric	M Metacentric

SM: arm ratio = 1.50–2.99. Centromere index = 40.0–25.1.
M: arm ratio = 1.00–1.49. Centromere index = 50.0–40.1.
**Sexual chromosomes.*
***Autosomal chromosomes.*

Table 5.
Classification of chromosomes of the Culex quinquefasciatus *from some regions of Iraq by arm ratio and centromere index by Levan et al. [38].*

These results provide important genetic information for understanding the chromosomal structure of *C. quinquefasciatus* mosquito from some regions of Iraq. This work opens the ways toward the creation of useful in detail studies of the chromosomes in additional mosquito species.

5. Conclusion

While all mosquitoes need standing water to reproduce, some of the other species have evolved so specifically that they will only lay their eggs in natural or artificial containers. The larvae of most mosquito species hang suspended from the water surface because they need air to breath. An air tube, called a siphon, extends from the larva's posterior to the water surface. Pupae are also physically active and employ a rolling or tumbling action to escape to deeper water, which is why they are commonly referred to as "tumblers." The pupal stage lasts from 1.5 to 4 days, after which the pupa's skin splits along the back allowing the newly formed adult to slowly emerge and rest on the surface of the water. The present study shows that the chromosomes karyotyping of C. *quinquefasciatus* species consists of three pairs of chromosomes (i.e., 2n = 6). Certainly much of this information is of a descriptive sort and easy to obtain by present techniques. Life cycle and karyotype studies in additional species can be simply done and should be carried out in a standardized way.

Author details

Sahar Abd
Department of Biology, College of Science for Women, Baghdad University, Baghdad, Iraq

*Address all correspondence to: sahar_abd2008@yahoo.co.uk

References

[1] Russel TL, Govella NJ, Aziz S, Drakely CJ, Kachur SP, Killeen GF. Increase proportions of outdoor feeding among residual malaria vector population following increased use of insecticides treated nets in rural Tanzania. Malaria Journal. 2011;**10**:80

[2] Abo El Hab JK. Medical and Veterinary Insects in Iraq (Theoretical Section). Iraq, Baghdad: College of Agriculture, University of Baghdad; 1979. p. 450

[3] Lacey LA, Undeen AH. Microbial control of black flies and mosquitoes. Annual Review of Entomology. 1986;**31**:265-296

[4] Oduola AO, Awe OO. Behaviour preference of *Culex quinquefasciatus* in human host in Lagos metropolis Nigeria. Journal of Vector Borne Diseases. 2006;**43**:16-20

[5] Hubalak Z, Halouzka J. West Nile fever. A reemerging mosquito borne disease in Europe. Emerging Infectious Diseases. 1999;**5**(5):643-650

[6] Rozendal JA. Vector Control, Methods for Use by Individuals and Communities. Geneva: World Health Organization, Regional Office for the Middle East; 2004. p. 379

[7] Korzeniewski K. The epidemiological situation in Iraq. Przegl Epidemiol. 2006;**60**:845-855

[8] Rozendal JA. Vector Control, Methods for Use by Individuals and Communities. Geneva: World Health Organization; 1997. p. 412

[9] Moosa Kazemi SH, Karimian F, Davari B. Culicinae mosquitoes in Sanandaj country, Kurdistan province, Western Iran. Journal of Vector Borne Diseases. 2010;**47**:103-107

[10] Larrick S, Connelly R. Featured Creatures. Department of Entomology and Nematology, USA, Florida: University of Florida; 2009. Available from: http://www.entnemdept.ufl.edu/creatures/aquatic/southern-house – mosquito

[11] Harun RB. Studies on the mosquito fauna in an urban and suburban area in Penang and the laboratory efficacy of mosquito coils containing different active ingredients against selected vector mosquitoes [M Sc thesis]. Malaysia: University of Malaysia; 2007

[12] Eldridge BF, Edman JD. Medical Entomology. Davis, USA: Department of Entomology, Center for Vector-Borne, Disease Research, University of California; 2000. p. 659

[13] Foote RH, Cook DR. Mosquitoes of Medical Importance. Washington D.C.: U.S. Government Printing Office; 1959 p. 160

[14] Service MM. Guide to Medical Insectology. Ali Mahmoud Sulait; Zuhair Mahmoud Al-Saffar and Riyad Ahmad Al-Iraqi (Translators). Directorate of the House of Books for Printing and Publishing, University of Mosul. 1984. p. 485

[15] Harbach RE. The mosquitoes of the subgenus *Culex* in Southwest Asia and Egypt (Diptera: Culicidae). Contributions of the American Entomological Institute. 2011;**24**(1):1-246

[16] Mullen G. Medical and Veterinary Entomology. 2nd ed. London: Academic Press; 2009. p. 637

[17] Hanson SM, Craig GB. *Aedes albopictus* (Diptera: Culicidae) eggs: Field survivorship during Indiana winters. Journal of Medical Entomology. 1995;**32**(5):599-604

[18] Adeleke MA, Mafiana CF, Idowa AB, Adekunle MF, Dansu BM. Morphometric studies on *Culex quinquefasciatus* and *Manson africana* (Diptera: Culicidae) in Abeokuta, South Western Nigeria. Tanzania Journal of Health Research. 2008;**10**(2):99-102

[19] Yadav P, Foster WA, Mitsch WJ, Grewal PS. Factors affecting mosquito populations in created wetland in urban landscapes. Urban Ecosystems. 2012;**15**(2):499-511

[20] Zinser M, Ramberg F, Willott E. *Culex quinquefasciatus* (Diptera: Culicidae) as apotential West Nile virus in Tucson, Arizona: Blood meal analysis indicates feeding on both humans and birds. Journal of Insect Science. 2004;**4**(20):1-3

[21] Sarwar M. Mosquitoes (Diptera: Culicidae) as malaria transmitters and procedures for suppression to exposure and spread of vectors. Biomedical and Health Informatics. 2016;**1**(2):38-43

[22] Sarwar M. Insect vectors involving in mechanical transmission of human pathogens for serious diseases. International Journal of Bioinformatics and Biomedical Engineering. 2015;**1**(3):300-306

[23] Sarwar M. Direct possessions of insect arthropods on humans owing to allergen, bloodsucking, biting, envenomation and stinging side by side case diagnosis and treating. International Journal of Bioinformatics and Biomedical Engineering. 2015;**1**(3):331-337

[24] Sarwar M. Insect borne diseases transmitted by some important vectors of class Insecta hurtling public health. International Journal of Bioinformatics and Biomedical Engineering. 2015;**1**(3):311-317

[25] Sarwar M. Problem created owing to insects in carrying vector borne diseases and combined vector control approach. International Journal of Chemical and Biomolecular Science. 2015;**1**(4):303-309

[26] Sarwar M. Skin disorders inflicted through insect invertebrates along with diagnosis and treating of cases. Journal of Nanoscience and Nanoengineering. 2015;**1**(4):233-240

[27] Sarwar M. Proposals for the control of principal dengue fever virus transmitter *Aedes aegypti* (Linnaeus) mosquito (Diptera: Culicidae). Journal of Ecology and Environmental Sciences. 2014;**2**(2):24

[28] Sarwar M. Defeating malaria with preventative treatment of disease and deterrent measures against Anopheline vectors (Diptera: Culicidae). Journal of Pharmacological and Toxicological Studies. 2014;**2**(4):1-6

[29] Sarwar M. Proposing solutions for the control of dengue fever virus carrying mosquitoes (Diptera: Culicidae) *Aedes aegypti* (Linnaeus) and *Aedes albopictus* (Skuse). Journal of Pharmacological and Toxicological Studies. 2014;**2**(1):1-6

[30] Sarwar M. Role of secondary dengue vector mosquito *Aedes albopictus* Skuse (Diptera: Culicidae) for dengue virus transmission and its coping. International Journal of Animal Biology. 2015;**1**(5):219-224

[31] Sarwar M. Reducing dengue fever through biological control of disease carrier *Aedes* mosquitoes (Diptera: Culicidae). International Journal of Preventive Medicine Research. 2015;**1**(3):161-166

[32] Sarwar M. Stopping breeding of dengue virus spreader *Aedes* mosquitoes (Diptera: Culicidae) with environmental modifications. International Journal of Bioinformatics and Biomedical Engineering. 2015;**1**(2):169-174

[33] Sarwar M. Elimination of dengue by control of *Aedes* vector mosquitoes (Diptera: Culicidae) utilizing copepods (Copepoda: Cyclopidae). International Journal of Bioinformatics and Biomedical Engineering. 2015;**1**(1):53-58

[34] Sarwar M. Controlling dengue spreading *Aedes* mosquitoes (Diptera: Culicidae) using ecological services by frogs, toads and tadpoles (Anura) as predators. American Journal of Clinical Neurology and Neurosurgery. 2015;**1**(1):18-24

[35] Sarwar M. Intervention focused on habitat modifications for ending up the *Anopheles* mosquitoes implicating in malaria transmission. American Journal of Clinical Neurology and Neurosurgery. 2015;**1**(2):126-132

[36] Sarwar M. Source reduction practices for mosquitoes (Diptera) management to prevent dengue, malaria and other Arboborne diseases. American Journal of Clinical Neurology and Neurosurgery. 2015;**1**(2):110-116

[37] Sarwar M. Mosquito-borne viral infections and diseases among persons and interfering with the vector activities. International Journal of Vaccines and Vaccination. 2016;**3**(2):00063

[38] Levan A, Fredga K, Sandberg AA. Nomenclature for centromeric position on chromosomes. Hereditas. 1964;**52**(2):201-220

Chapter 4

The Yellow Fever Mosquito *Aedes aegypti* (Linnaeus): The Breeding Sites

Márcia Ramos Jorge, Antonio Pancrácio de Souza, Ricardo Augusto dos Passos, Silvia Maria Martelli, Chaiane Regina Rech, Alexeia Barufatti, Bruno do Amaral Crispim, Helina dos Santos Nascimento and Eduardo José de Arruda

Abstract

Important information about yellow fever mosquito *Aedes aegypti* (Diptera: Culicidae) and the identification and quantification of the main Brazilian breeding sites for this vector are highlighted in this chapter. Although most of the control actions have been directed to the adult (winged) phase of the vector, the reduction of immature forms (eggs and larvae) of *Aedes* is the most important way to control these insects, especially to eliminate breeding sites. These are, in principle, the most important targets for mosquito population control in order to reduce infestation and, consequently, the transmission and incidence of diseases transmitted by insect vectors. Thus, this chapter presents a compilation and discussion that allows comparing characteristics and similarities of *Aedes* species.

Keywords: mosquito vectors, population control, *Aedes aegypti*, dengue, public health

1. Introduction

Mosquitoes in the family Culicidae are responsible for the transmission of viruses and other pathogens to man and animals. Arboviruses transmitted by mosquitoes of the genus *Aedes* (Meigen) (Diptera: Culicidae) promote the occurrence of diseases such as dengue fever, chikungunya fever, yellow fever and zika fever at alarming levels and have impacted public health in several countries, mainly Brazil [1, 2].

In Brazil, there is an outbreak of wild yellow fever [3] due to virus transmission, which should be kept in the wild, between non-human primates and arboreal wild mosquitoes, mainly of the genera *Haemagogus* (Linnaeus) (Diptera: Culicidae) and *Sabethes* (Robineau-Desvoidy) (Diptera: Culicidae). The possibility of a change in the ongoing yellow fever transmission cycle has been realized. To date, it has not been proven that the yellow fever mosquito *Aedes aegypti* (Linnaeus) (Diptera: Culicidae) mosquito is involved in transmission, but evidence has emerged from different studies [3].

With the exception of yellow fever and dengue, zika and chikungunya do not have specific vaccines or treatments. For this reason, the use of insecticides and chemical control constitutes the main insect vector control tools in urban areas. However, this practice has been disseminated and performed irresponsibly, resulting in actions that caused resistance to these insecticides. In this respect, new forms of control are desirable, especially with multifunctional products [3].

In Brazil, *Ae. aegypti* is found in more than 70% of the 5561 Brazilian municipalities. Most of these areas have a reduced number of health agents who, for different reasons, do not cover 80% of cities' urban areas. Thus, the country is left exposed to constant threats of outbreaks and epidemics. Thereby, it is necessary to develop new products, methodologies or even a combination of methods to control *Aedes* spp., populations. These methods should take into account effective surveillance and control, and be of a low cost and easy applicability. Additionally, efforts to improve infrastructure, population education, and training of health workers to combat and control the insect should be intensified. To this end, a better understanding of the different aspects of Culicidae biology and diversity, their relationships with humans and animals, and their reproduction in urban areas are important factors in assessing the adequacy of methodologies used for the selection and implementation of control methods, the development of new insecticides or the creation of new strategies that consider behavioral and adaptive aspects. However, information, techniques and/or scientific knowledge are still restricted to the majority of the population or have limited practical applications. It is up to the universities, research centers and zoonotic control agencies to obtain them and use the contents for discussion or establishment of new proposals and strategies of vector population control.

In this context, it is very useful to present a review of breeding sites, habits and behaviors of the *Aedes* species to understand the situations, conditions and/or details of the microenvironments, and reproduction at breeding sites, whether natural or artificial. Thus, this chapter is based on the collection and analysis of information available in the literature, or the authors' scientific experience, about the physicochemical and biological conditions of breeding sites and mosquito habits and behaviors. Considering these important findings will crucially enable the elaboration of strategies that control insect vector populations.

2. Reproductive cycle, behaviors and habitats

Mosquitoes use the aquatic environment to develop immature forms (egg, larva and pupa). In urban and peri-urban environments, breeding sites are any artificial reservoirs or containers and even natural breeding grounds where water is available for the reproductive process of the species. The availability of breeding sites and feeding facilities explains, in part, the spread of *Aedes* spp. *Ae. aegypti* is considered the main vector of (re)emerging diseases in Brazil (yellow fever, dengue, chikungunya and zika) and has been a constant concern for public health.

The genus *Aedes* [including *Ae. aegypti* and Asian tiger mosquito *Aedes albopictus* (Skuse) (Diptera: Culicidae)] has peculiar habits in relation to other mosquito species. The species oviposit in a split way, usually in aquatic environments with areas of restricted lighting or shaded spaces. At these locations, temperatures are milder compared to an area of direct solar sunshine. Shaded spaces and the availability of organic debris are physical and chemical signs that the breeding facility was suitable for previous generations and can successfully serve the breeding of a new generation [4].

Breeding sites are important spaces for the attraction, reproduction, survival guarantee and continuation of the species. These microenvironments are determinants for oviposition and reproductive success. Food availability in the aquatic environment appears to be the main attraction for oviposition, along with the physicochemical and biological conditions of the breeding sites. These related aspects can all influence the choice of the oviposition site by the females.

The abundance and population density of mosquitoes produce negative impacts on the quality of life of humans and animals. Diseases caused by arthropod-borne arboviruses are considered as neglected or emerging diseases. Viral diseases transmitted by *Aedes* spp., including yellow fever, dengue fever, chikungunya and zika, are serious public health problems that persist for decades in the form of periodic outbreaks with variable intensities due to non-continuous and deficient control methods. Since the early 1970s, the World Health Organization (WHO) has sought an effective way to control, manage and treat these diseases, but without much success.

Ae. aegypti undergoes four stages in its life cycle (**Figure 1**): egg (terrestrial form), larva, pupa (aquatic forms) and adult insect (winged; terrestrial form). Despite having a short shelf-life of at most 2 months in the laboratory (and less time in natural environments), the female can oviposit up to 270 eggs during her life span. As a strategy for species survival, egg laying and establishing an egg bank at breeding sites allows the maintenance, replacement and perpetuation of the species even under adverse conditions [5]. Split oviposition at the breeding sites performed by *Ae. aegypti* and *Ae. albopictus* allows them to increase survival conditions and establish the population in the ecosystem.

Rozeboom et al. [6] observed that females disperse their eggs through several breeding sites. In each site, 11–30 eggs are deposited by oviposition in small breeding sites and more than 60 eggs in large breeding sites [7]. Vector control strategies should include knowledge about the reproductive aspects of the insect due to the formation of an egg bank that allows for species reposition in the environment. Studies showed that both preferences and habits, as well as insect survival strategies, are important factors that must be considered by the vector control agent [7, 8].

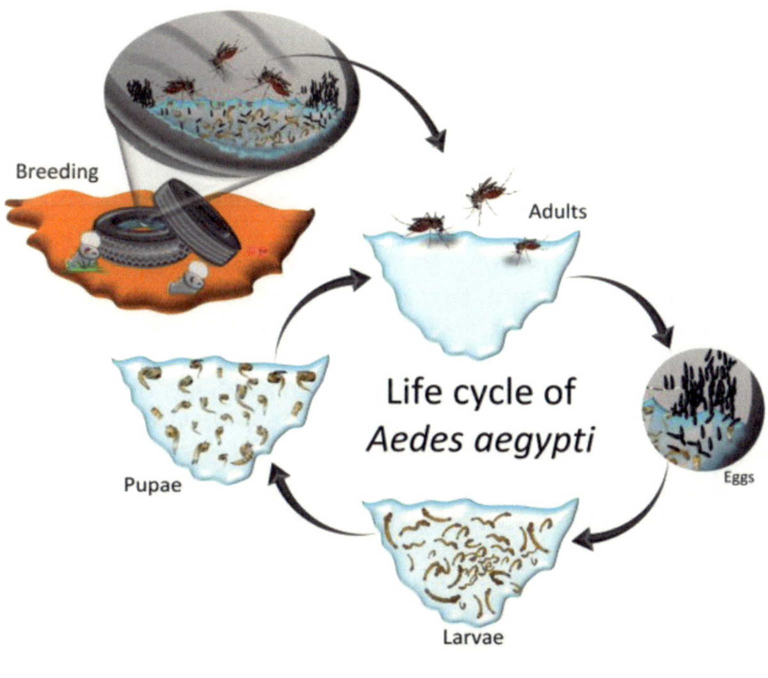

Figure 1.
Life cycle of Aedes aegypti *(Linnaeus) (Diptera: Culicidae).*

The literature mentions that the breeding site must be viable for reproduction of the species, with food availability and adequate oxygenation, temperature and pH. The choice of oviposition sites is a critical factor for the survival and dynamics of populations, and knowledge of these factors is important for mosquito population control. The choice of such sites results in a complex interaction between chemical and physical factors (biotic and abiotic), which indicates the degree of preference in the selection of breeding and oviposition sites [4–9]. Laboratory studies done by Madeira et al. [10] showed *Ae. aegypti* preferences for some types of breeding grounds. There are variations in the type of breeding sites chosen by this species for oviposition. Oviposition can occur in the water when the egg-laying surface does not allow egg adhesion. Non-submerged eggs usually do not hatch, in contrast to submerged eggs that hatch after hydration stimulation. Mosquitoes prefer porous or rough surfaces for oviposition.

Mitchell-Foster et al. [11] showed that new strategies can be used for integrated vector control. The imposition of unfavorable conditions for insects at breeding sites can help to increase insecticide effectiveness and vector control, since insecticides can be multifunctional for comprehensive secondary or tertiary control, due to changes in breeding sites physicochemical and/or microbiological conditions. These strategies could be integrated to consider applying disadvantageous situations and they include biological agents at different levels for *in situ* reproduction control. Exposure to competitors, physicochemical changes and others reduce the probability of egg hatching and species survival, besides affecting parameters such as larval maturation time, adult size, population density, fecundity, attraction and egg viability.

Rey and O'Connell [12] investigated *Ae. aegypti* and *Ae. albopictus* oviposition behavior by examining oviposition site characteristics and the influence of factors that alter egg laying. In cages, they prepared several combinations with females of both species for oviposition at breeding sites in vegetation (potted plants) or structural components (blocks of wood and concrete). Then they compared the number of eggs deposited per female among the species and the different types of breeding sites. *Ae. aegypti* deposited more eggs per female in breeding sites with vegetation compared to those with structural elements. The opposite is true for *Ae. albopictus*. *Ae. aegypti* females presented a higher oviposition frequency than *Ae. albopictus*. On average, 63% of breeding sites contained eggs from both species. The authors emphasized the possibilities of interactions between *Ae. aegypti* and *Ae. albopictus* in spaces and breeding grounds, especially when the two species are disseminated in the same space where specific interactions and mechanisms must contribute to reproductive cooperation.

After oviposition on the breeding walls, white eggs in the posture darken on contact with atmospheric oxygen. Eggs are deposited on the walls of the breeding site when the surfaces are porous or rough and close to the water surface (but not directly over the liquid when there is this availability of an oviposition surface). Humidity must be maintained for egg viability and hatching after submersion/hydration in water. Studies revealed that these eggs can survive 360–450 days after desiccation. Under favorable conditions, their hatching can occur between 10 minutes and 48 hours after humidification or immersion in breeding liquids, and after 7–10 days, they emerge as adult insects with vector potential, depending on the temperature and available feeding conditions [13, 14].

Mosquitoes seek shelter, dark spaces with microbiota and decomposing organic wastes available over a wide pH range (acidic, neutral and basic). If the insect does not find these conditions, it will look for other places, regardless of the size or condition, for their reproduction. The insect will never put all its eggs in a single breeding site. This survival strategy led to marked reproductive success and greater dispersion, features that make the mosquito resilient, competent and adapted to the different conditions offered by the medium for its survival, despite its apparent fragility. Its adaptive and vectoral capacities underscore its epidemic potential due to the ability to spread across different geographical areas [15].

After hydration and egg hatching, the larval stage begins, which has four evolutionary steps. Its duration depends on temperature, food availability (organic material or microorganisms) and number of larvae in the breeding site. Under these conditions, the period between hatching and the pupa (last stage before the adult insect) may not exceed 5 days. However, if there is low temperature and food shortage, the stage can be extended for up to a week [14].

The *Ae. aegypti* larva is composed of a head, thorax and abdomen. The abdomen is divided into eight segments. The posterior and anal segment of the abdomen has four lobulated bronchi for osmotic regulation and a siphon or air tube for breathing on the water surface. The siphon is short, thick and darker than the body. To breathe, the larva goes to the surface, where it is in almost vertical position. It moves in the shape of a serpent, making an "S" in its displacement. It is sensitive to sudden movements in the water and under a beam of light, it moves quickly, seeking refuge at the bottom of the vessel. This behavior is more related to antipredation because females select the oviposition site based on the success of previous generations as well as the presence of decomposing organic debris [4].

The pupae do not feed and at this stage metamorphosis of the larval stage occurs to generate the adult (winged) insect. This stage lasts 2–3 days and when the pupa is inactive it stays floating on the water surface to facilitate the emergence of the adult (winged) insect. The pupa is divided into the cephalothorax and abdomen. The head and thorax are united and constitute the cephalothorax, which gives the pupa, seen from the side, the appearance of a comma. The pupa has a pair of breathing tubes or "trumpets" that cross the water and allow it to breathe [14, 15].

The adult phase of the insect represents the reproductive and transmitter/vector phase. The adult is responsible for the spread of viruses and diseases and for the reproduction. *Ae. aegypti* is approximately 1 cm long and have black with white stripes distributed near the body and legs. *Ae. aegypti* incidence is usually greater in spaces with high temperature, low humidity, low vegetation and a concentration of heat sources. Unlike common mosquitoes, *Ae. aegypti* has diurnal habits, flies low and performs hematophagy, preferentially stinging the feet, ankles, arms and legs. *Ae. aegypti* does not like heat or radiated/lighted environments, so it is more active in the early morning and late afternoon. It is believed that the environment in these adverse conditions of irradiation and/or competitiveness with other species is more aggressive for the adult mosquito, which imposes an additional metabolic cost to its survival. The female feeds on blood (hematophagy), which is a ready and metabolized food source that is still rich in proteins, and minerals along with other micronutrients necessary for its survival, oviposition and the maintenance of its vector competence. However, the female only transmits a virus if she contracted it more than 2 weeks ago, since the dengue virus needs 10–14 days in mosquito metabolism to become viable for transmission during hematophagy [16].

2.1 The breeding sites of *Aedes*

The availability of different types of discarded containers with organic remains allows rainwater containment. Maintaining these conditions for a certain period of time allows the formation of new breeding sites that attract females for oviposition and ultimately promotes Aedes larval infestation.

Reproductive habitat characteristics may affect larval population density. Thus, the mosquito will be more abundant and the potential for disease transmission will increase. In this situation, quality of life is deeply affected by vector diseases, especially for individuals who live in high-population-density areas where disease will be transmitted more quickly.

Getachew et al. [17] showed the importance of basic sanitation and the simple treatment of water with chlorine, which represents an important control for the restriction of disease outbreaks. In containers that contain tap water, mosquito larvae are not abundant, contrary to what happens in containers filled with a mixture of tap water and rainwater or only rainwater. De Brito-Arduíno and Ávila [18] researched and discussed the reproduction and characteristics of urban breeding sites, and the implications for *Ae. aegypti* and *Ae. albopictus* control. The authors' results corroborate studies performed in Australia, where 28% of immature individuals came from tanks and wells, considered to be the main breeding sites [19]. Other studies performed in Venezuelan cemeteries showed that the breeding sites with the highest proportion of pupae have been those that containing 1–5 liters of liquid [20]. Additionally, the results presented great variability in the parameters, since immature individuals of *Aedes* spp., occur in water with pollutant load [19].

The species can develop in breeding grounds with a high degree of pollution and organic matter. The polluted breeding sites found in De Brito-Arduino and Ávila's studies [18] are boats, drains and sites that contained the residues of various substances, such as oil, salt, rust, among others [21]. In studies performed in Puerto Rico, a large number of *Ae. aegypti* larvae grew in tanks and septic tanks, cooperatively with *Culex quinquefasciatus* (Say) (Diptera: Culicidae) populations [22, 23]. In Brazil, laboratory analysis showed that *Ae. aegypti* can complete its reproductive cycle up to the adult (winged insect) phase in waters with different degrees of pollution [21–24]. In Merida, Mexico, the main *Ae. aegypti* breeding sites are composed of rainwater runoff from the streets and localities that have a high pollutant load [25].

Ae. aegypti prefers breeding sites with high food availability when there are no conspecifics (*Ae. albopictus*), while *Ae. albopictus* prefers high food availability when larval density is relatively low [25, 26]. The literature suggests that *Ae. aegypti* possesses the ability, genetics and competence to adapt and survive in different biotic and abiotic conditions that result from anthropogenic changes. Studies revealed that larvae and pupae are found in breeding grounds with water of different quality and portability. Some breeding sites contain residues of different substances, as well as different temperatures, pH, dissolved oxygen (DO) and conductivity. These results demonstrate that the species can survive in non-conventional breeding sites [27–29].

Taken together, these facts highlight the challenge for mosquito population control and the reduction of viral disease incidence in urban and peri-urban environments. Effective *Ae. aegypti* population control to reduce disease incidence should focus on multifactorial strategies and take into account the different physicochemical and biological aspects of breeding sites.

2.2 Physical-chemical characteristics of breeding sites

The physicochemical characteristics of aquatic habitats influence the survival rate of mosquitoes [15–17]. Some factors may be important for mosquito oviposition, including water temperature, pH, ammonia, nitrate, sulfate and phosphate content, conductivity and dissolved solids [30, 31].

Following oviposition on the breeding wall, eggs hatch after hydration, and the insects develop through the four larval stages and pupa (which does not feed) until adulthood. The immature stages, namely eggs and larvae, are important parts to consider in population control. From egg hatching (embryo), the larvae consume decomposing organic material, debris and microorganisms from the aquatic microenvironment to obtain nutrients for their development, which are actions that contribute to reproductive cycle closure. These aspects can and should be considered for the establishment of new insecticides, formulations and strategies for insect population control. Control via the food chain, studies on microbiota composition (including prevalent and non-prevalent species), mosquito characteristics and specificities, the attractiveness of the breeding grounds to females and other details are important considerations to develop products and strategies for target species population reduction. Current research reveals that the type of container, water quality and the available physical, chemical, biological and nutritional conditions found at the breeding microenvironment are important parameters for the attraction, reproduction and continuation of the species. Thus, these factors are likely imperative for population control.

According to the historical incidence of vector-borne diseases, infection peaks occur during hot and rainy seasons. At these times, water storage lasts for prolonged periods, with the potent combination of humidity, temperature and food availability. The hot and humid climate increases mosquito population density from these breeding grounds, food and conditions favorable for reproduction of *Aedes* spp., all of which increase the risk of the incidence of the different diseases transmitted by them [17].

Studies in Africa, Asia and various parts of the world indicate that the predominance of *Aedes* spp. (*Ae. aegypti* and *Ae. albopictus*) in a given region depends on the climatic conditions, type of breeding grounds and form of species dispersal. The use of abiotic parameters like water conductivity (salt dispersion/electrolytes) as an index for mosquito proliferation in garbage dumps can be explored in mosquito control strategies [30].

Getachew et al. [17] observed that water conductivity in the range of 162.9–616.9 $\mu S\ cm^{-1}$ negatively correlates with larval density. Fillinger et al. [32] also showed a negative correlation of larval density to malaria vector. A conductivity of 2000 $\mu S\ cm^{-1}$ significantly reduces larval density at breeding sites. The authors also showed that the breeding site pH ranged from 6.72 to 7.63, with a positive correlation with larval density.

Breeding site ionic composition and pH are limiting factors for the distribution and number of organisms found in these habitats. However, Clark et al. [33] showed that the mosquito has a high adaptation capacity to the breeding site conditions. *Aedes* spp. can develop in extreme pH (4.3–9.9), wide temperature ranges (18–35.9°C) and different salinity levels. The insects prefer environments with leaves and decomposing organic material, although it is not an absolute rule for *Aedes* spp., with reproduction capacity in waters polluted by oil, salt and rust, among other factors. Indeed, De Brito-Arduino and Ávila [18] observed breeding sites in boats with motor oil residues, data that may also justify the minimum dissolved oxygen (DO) value of 0.40 $mg\ L^{-1}$. The drains also had low DO (0.20–0.52 $mg\ L^{-1}$) in the various collected samples, besides the presence of soap residue and detergents that would justify these DO values.

Another factor that influences the choice of breeding sites is salinity. Oviposition decreases with increasing salinity at concentrations above 12% [17, 30, 32, 34]. De Brito-Arduíno and Ávila [18] showed that these limits are flexible, with salinity values up to 13.5%. Increases in conductivity (salinity) impose unfavorable conditions for oviposition [18, 35]. In the littoral strip of Brazil and Asia, there are records of *Ae. aegypti* reproducing in water with high salinity (brackish water), data that demonstrate the existence of the species adaptation to the salinity of the breeding sites [21, 24].

Another study showed that breeding sites can be shared by different species for survival at different pH ranges. *Ae. aegypti* and *Cx. quinquefasciatus* (common stilt) have been found in breeding sites with pH in the range of 6.5–8 [36]. Brito-Arduino and Ávila [18] also observed that the breeding pH range varies from acidic to basic, with values from 4.3 to 9.9. These results are related to other studies performed in Australia, which showed a breeding site pH range of 5.5–11.3 [37], and in Venezuela, with pH between 6.5 and 10.5 [20]. Since the hydrogen ion concentration [38] is dependent on the water containment medium (breeding sites) and temperature, due to the water ionization constant, temperature is another factor that deserves careful analysis.

Temperature influences the immature development of *Ae. aegypti*, and thus, the availability of energy from the medium may interfere with metabolic activity, but mainly with the enzymatic activities of the organism. High temperatures can compromise the conformation, stability and function of insect metabolic enzymes, as well as activate and deactivate them or even alter metabolic processes that are involved in the development, adaptability and survival of these organisms [33, 37, 39]. As previously stated, the species can successfully develop and breed between 18 and 35.9°C [18]. Studies performed in Colombia revealed that *Ae. aegypti* larvae could grow from 24 to 30°C at breeding sites [39, 40]. Another related study conducted in Nukulaelae (Tuvulu), with breeding grounds in metal/plastic drums, showed that *Ae. aegypti* occur in environments with temperatures between 20 and 35°C [41]. The pH and salinity may limit the distribution of organisms. However, they cannot prevent reproduction under unfavorable abiotic conditions. The conductivity (salinity) of the water in the breeding sites of De Brito-Arduíno and Ávila [18] study, ranged from 0 to 22,400 μS cm^{-1} (0–13.5% salinity) and pH from 4.5 to 9.9 (depending on the collection and type of breeding sites). These results suggest that *Ae. aegypti* can reproduce under unfavorable conditions and thus overcome limiting factors for species development in extreme temperature, pH, DO and conductivity conditions.

In a study conducted by Rao et al. [30] in *Ae. albopictus* breeding sites, there is a positive correlation between larval density and total dissolved solids, turbidity, conductivity and salinity, and a negative correlation for pH. Dom et al. [42] and Nazri et al. [43] showed the potential risk of mosquito peri-domiciliary establishment and the characteristics of the breeding sites for *Aedes* spp. *Ae. albopictus* is the predominant species in breeding sites. Turbidity, pH, total organic carbon (TOC), magnesium, calcium and sodium are among the characteristics that are positively significant for larval density, species composition and breeding water characteristics, respectively. Mosquito breeding sites characteristics affect larval density and increase mosquito infestation, which ultimately impact the quality of life of the population.

Breeding characteristics play an important role in providing nutrients and a safe environment for increasing the larval survival rate. Different studies showed that turbidity (due to the presence of organic material and microorganisms), pH, TOC and magnesium concentration influence the larval population density. *Ae. aegypti* prefers less turbid water, while *Ae. albopictus* prefers more turbid water. These differences may indicate why one species prevails over another. The results suggest that *Ae. aegypti* is a less-demanding species in terms of food for oviposition and reproduction.

TOC, related to the presence of organic material in breeding sites, also influences larval density. This fact has been confirmed by a study of Thangamathi et al. [44], which indicated that sulfate ion (so_4^{-2}) content in the breeding habitat of *Aedes* larvae (*Ae. aegypti* and *Ae. albopictus*) is high compared to other ionic species. *Aedes* spp. have a rapid adaptation capacity to environmental conditions, even if they appear to be averse to the species. According to Thangamathi et al. [44], *Ae. aegypti* survives most at pH between 6.5 and 8.0; however, values close to neutrality are better for the species [45]. This pH range may be altered to affect species adaptation and potentially to control the population.

Micro and macro nutrients availability in breeding sites is also important for *Ae. aegypti* and *Ae. albopictus*. Magnesium plays a role in larval density, as shown by Thangamathi et al. [44], who carried out studies that analyzed breeding site mineral composition. Calcium, potassium, sodium, sulfur (elemental or SO_4^{-2}) and magnesium [2] can provide an ideal broth for *Ae. aegypti*. The larval habitat characteristics have been analyzed and related to the species by larval number. Calcium and sodium provide significantly different results for the proposed composition that favors *Aedes* spp. in breeding sites.

These results show that any container that can store water for a prolonged period for domestic or industrial use becomes a potential mosquito breeding ground, since it can allow the deposition of minerals and organic material, colonization by microalgae and microorganisms, and thus gather conditions for the development of a microenvironment favorable to the life and reproduction of *Aedes* spp. In this regard, strategies such as insecticides or other products could potentially change breeding site physicochemical parameters in terms of pH, conductivity and nutrient availability for secondary vector control (due to reproductive microenvironment changes).

2.3 Attractiveness of females to breeding sites

The *Ae. aegypti* oviposition strategy differs from other mosquito species. The literature shows that the oviposition stimulus is mainly food availability, signaled by microbiota semiochemicals that result from the degradation of organic materials and/or microorganisms. Semiochemicals are signaling molecules, used to carry information between living organisms that cause changes in *Ae. aegypti* behavior. These semiochemicals are perceived by the female, stimulate oviposition and consequently, allow species continuation, possibly because they indicate food availability for the larvae.

There are few studies that relate the attractiveness, selectivity and the relationship between microorganisms (i.e., bacteria) and products of organic degradation of breeding sites for attraction and oviposition. The available reports pay little attention to the density and colonization of breeding grounds by bacteria and other organisms and the female's attraction or repulsion and oviposition. Hasselschwert and Rockett [46] examined *Culex* larvae growth at breeding sites with different bacterial cultures and observed that the presence of *Bacillus cereus* and *Pseudomonas aeruginosa* induces attraction for *Ae. aegypti* oviposition. Trexler et al. [47] showed that isolates of *Sphingobacterium multivorum* (Sphingobacteriales: Sphingobacteriaceae), *Psychrobacter immobilis* (Juni and Heym) (Pseudomonadales: Moraxellaceae) and an unidentified *Bacillus* (Cohn) (Bacillales: Bacillaceae) species, when inoculated in water, induce high *Ae. albopictus* attraction and oviposition compared to water without these bacteria [40]. Poonam et al. [48] produced filtrates from bacterial cultures and used them for analysis of attraction and oviposition with female *Cx. quinquefasciatus*. These females oviposited more in certain concentrations of *B. cereus* (Frankland and Frankland), *Bacillus thuringiensis* (Berliner) and *Pseudomonas fluorescens* (Flügge) (Pseudomonadales: Pseudomonadaceae) compared to tap water. Other filtrates from *Bacillus megaterium* (Bary) and *Azospirillum brasilense* (Tarrand, Krieg and Döbereiner) (Rhodospirillales: Rhodospirillaceae) did not stimulate oviposition in water at any concentration of the filtrate tested. These results demonstrate that there is selectivity in the attraction and response of the mosquito to different bacteria. Bacteria grown in disparate nutritional media and this medium is not used as a control because it may contain substances that are attractive or deterrents for oviposition.

Huang et al. [49], in a study with 14 isolates of bamboo leaf infusion bacteria, showed that *Ae. aegypti* are repelled at densities of 109 cells mL^{-1}. The mixture of 14 bacterial isolates attracts *Ae. aegypti* and *Ae. albopictus* at densities of 107–108 cells mL^{-1}. These studies showed a concentration range dependence of 106–109 cells mL^{-1} for *Aedes* spp. attraction. *Ae. aegypti* and *Ae. albopictus* females are attracted by an infusion of bamboo leaves and a mixture of 14 bacterial species [50–52]. The results indicate that volatile substances from plants or bacteria can be attractive to mosquitoes. Despite the differences in the responses of *Ae. aegypti* and *Ae. albopictus* to different bacterial concentrations and densities, *Ae. aegypti* females tend to be attracted to lower bacterial concentrations and densities compared to *Ae. albopictus* females.

Other data available in the literature indicate that not all bacterial species produce the same semiochemicals (attractive or repellent) or sufficient amounts of attractants. These results suggest that there is a possible biological balance in the microbiota that controls, from the population density of the present species, the emission of volatile substances that can be perceived by the females in the process of attraction or repulsion to the breeding sites. Semiochemical composition variations from different bacterial species that stimulate the attraction or repulsion and oviposition of *Aedes* spp. explain the different species response. Further, other parameters, including variations in bacterial production time, can promote quantitative or qualitative differences for the chemical stimulus in the process of attraction or repulsion to the breeding sites and humans skins emanations [49, 53]. Thorn et al. [54] described significant differences in the type and concentration of volatiles produced by different bacteria grown in the laboratory and in nature for oviposition [6]. Ponnusamy et al. [50] showed the complexity of a microbiota balance and its importance in breeding sites for mosquito breeding. Their studies reinforce the hypothesis that the main stimulus for female ovipositors is the availability of food indicators or breeding site microbiota.

2.4 Perspectives of integrated pest management (IPM) applied to population control of vectors

Diseases transmitted by *Ae. aegypti* are one of the major public health problems in many regions of the world; hence, their control is import. Thousands of years of natural selection increased vectoral competence and even species survival to the cultural habits of populations/communities. Additionally, lack of health education, high availability of breeding sites in domiciles and peri-domestic areas, and the absence in many places of sanitary infrastructure, among other factors, contribute to increase the survival and reproduction success of mosquito species and prevent their population control.

Despite of reports and dissemination of the polyvalent vaccine for dengue, chikungunya and zika that may be an option to reduce and control these diseases, this use can take up to 10 years for successful control, according to the Oswaldo Cruz Foundation (Fiocruz, RJ), due to the conditions required to guarantee efficiency and biosafety. There are four dengue serotypes and the number of serotypes for the other two viruses is still unknown, and must also be combated efficiently. For this reason, there is enormous dependence on chemical control and the need for eliminating *Ae. aegypti* breeding sites to reduce the mosquito population and consequently, the incidence of the diseases. The vaccine may be an important part for dengue control; however, other fronts should also be considered, such as health education, continuous epidemiological surveillance, infrastructure improvements and social responsibility by governments.

2.4.1 Mechanisms of resistance to chemical insecticides

Chemical methods are intensively used for adult insect vector population control. Non-implementation of preventive control strategies is in part, responsible for the occurrence and constant aggravation of epidemic outbreaks during hot periods in Brazil.

The difficulty for *Ae. aegypti* population control centers are on the adaptive process, species resilience and acquired resistance to conventional insecticides. The first report of resistance occurred with organophosphates followed by pyrethroids. Over time, other resistance records have been reported to organophosphates, pyrethroids and carbamates in regions with intense mosquito occurrence. Before the year 2000, resistance to the organophosphate insecticides Temephos used as a larvicide and Malathion used to combat adult mosquitoes, was detected in some Brazilian municipalities.

Reproductive capacity favors the emergence of insecticide-resistant individuals used for vector control, because resistance varies with the time of use and concentration of the products that act at specific sites of toxicity. Medeiros [55] presents four categories of enzymatic activity profiles: (1) greater activities in the adult stage; (2) greater activities in the larval stage (esterases "α-EST" and "β-EST"); (3) activities that increase during each stage evaluated (mixed function oxidase [MFO]); and (4) activities that tend to increase in the larval stage and decrease in the first days of adult life (DVA) [esterase "ρNPA" and glutathione S-transferase (GST)]. Biochemical assays with larvae and adults from field populations revealed alterations in acetylcholine esterase (AChE) and other esterases at the larval stage, changes in GST more restricted to the adult stage, and MFO alterations have been limited to the two vector stages.

The results from these experiments allow detailed evaluation of the resistance mechanisms in different vector populations and can be used for the development and choice of insecticides more suitable for *Aedes* spp. control. Guirado and Bicudo [56] showed some aspects of population control and resistance to insecticides in *Ae. aegypti* and concluded that mosquito population control, as long as no more modern vaccines or genetic techniques of epidemiological control are available, is exclusively dependent on chemical control and human population awareness for breeding site elimination.

The available insecticide resistance studies demonstrate that resistance is due to three main mechanisms: (a) reduction of insecticide penetration due to changes in the insect cuticle; (b) increased metabolism of the insecticide by the action of esterases, mono-oxygenases or glutathione-transferases that inactivate the molecule; and (c) modification of the insecticide's biological target. The literature also shows a behavioral resistance mechanism, where insects avoid contact with sites that contain the toxic substance used for control. The results of insecticide resistance and loss of control efficacy indicate the need for continuous monitoring of *Ae. aegypti* susceptibility to insecticides and the use of chemical control in more rational ways. These smarter considerations include analysis of insect populations and their resistance, use of integrated management techniques and different methodologies and/or control products, in addition to continuous monitoring during all periods of the year by endemic control programs.

2.5 Attractiveness to breeding sites from water quality

The availability of water by precipitation or accumulation at home and breeding grounds is an important factor for the reproductive process and establishment of

Aedes spp. [57, 58]. Additionally, food availability is the most important factor for larval survival and development. Researchers have analyzed breeding site physicochemical conditions and water parameters. The studies reveal species adaptations and tolerance to the physicochemical conditions of breeding sites, besides the presence of other species that share breeding site spaces and cooperate with each other. Gil et al. [29], through collection and analysis of breeding water, showed a close relationship between the conditions of drinking and non-potable water composition, temperature, availability of organic material, waste, and reproduction and development of *Ae. aegypti*. Beserra et al. [26] showed that *Ae. aegypti* develops in different environments with organic material availability. The temperature, pH and water composition (total solids, total nitrogen, ammoniacal nitrogen, total dissolved phosphate and DO) are important variables for *Ae. aegypti* reproduction at egg, larval and pupal stages.

Silva and Silva [59] studied the influence of egg quiescence (interruption in development induced by low humidity) period in the life cycle of *Ae. aegypti* under laboratory conditions, in search of information that could improve control actions. Notably, the egg is the most resistant form of the biological cycle, and this feature allows mosquito development and resistance to climatic adversities. The experiments have been performed in a biological chamber, maintained at 28 ± 1°C, with relative humidity of 80 ± 5% and a 12-hour photoperiod. Quiescence is highly significant for larval hatching. Eggs from the same quiescence period had statistically different incubation periods and the larvae hatched in groups (defined by incubation). Indeed, in 99.8% of the cycles, the variation was determined by incubation.

Quiescence is a very important adaptation in the passive dispersion of *Ae. aegypti* because it enables the transport of resistant (or quiescent) eggs into all kinds of artifacts, such as used tires. Its non-destruction can signal a feasible mechanism to increase *Ae. aegypti* dispersion, a factor that makes entomological and/or epidemiological surveillance essential. This context gains another dimension when one adds the highly significant effect of quiescence on larval hatching. Quiescence periods of up to 720 days can occur, but the literature demonstrated that eggs are still viable up to 492 days. In some studies, the highest hatching rate (97.2%) occurs after 121 days quiescence [59, 60]. In another study, high hatching occurs after 180 days [23]. Among these quiescence periods, the 121 days period is the most favorable to *Ae. aegypti*, with significantly higher hatching rate and number of cycles (groups) than for other periods.

Other authors [32, 34, 36] also observed the same phenomenon attributing to different factors these prolonged periods of viability that include resistant, durable, inactive and residual eggs. Whatever the factor is, quiescence provoked by climatic variables, such as the decrease of humidity and temperature, increases *Ae. aegypti* life expectancy by enabling re-growth of the species. This behavior is relevant to the control of this mosquito, since the eggs adhered to container walls resist desiccation and can hatch when the water level rises.

In dechlorinated water at 26°C, there is a 92% *Ae. aegypti* larvae survival rate [60]. The *Ae. aegypti* larvae grown in raw sewage water show shorter developmental periods, probably due to the high concentrations of organic materials that serve as nutrients, although the authors hypothesized that the surface tension and high viscosity of the water are responsible. Further, the formation of material on the surface may makes it difficult for the larvae to obtain atmospheric oxygen. Beserra et al. [60] showed that *Ae. aegypti* can develop in polluted environments such as domestic sewage, where there is a high concentration of organic material and practically zero DO (0.12 mg L^{-1}), as well as in treated effluents, with carbonaceous material removed, as the post-treated effluent after anaerobic filter and in a polishing pond.

Notably, food availability is a decisive condition in the selection of the breeding site by the female, even in conditions considered unfavorable with regards to the available water quality, including high pollution levels. Beserra et al. [60] analyzed water quality and food availability for *Ae. aegypti* larvae. Water quality elongates the breeding period, with or without food and the medium turbidity decreases. Further, turbidity is an important parameter and low light penetrability in the aquatic environment is beneficial for the species, since *Ae. aegypti* larvae are photophobic. The *Ae. aegypti* is a competent vector for the transmission of flavivirus, which causes yellow fever and dengue fever. In this organism, there are at least 10 different rhodopsins. Mosquitoes have similar sets of rhodopsin and retinal organization, including the malaria vector *Anopheles gambiae* (Giles) (Diptera: Culicidae) that diverged from *Ae. aegypti* millions of years ago. Conservation of molecules within these visual systems implies that these species have visual processing capabilities similar to those of the common ancestor. Thus, these capabilities remain important for the suitability of existing species and their adaptability to environmental conditions. In *Ae. aegypti*, several mechanisms contribute to the ability of a photoreceptor to adapt to ambient light conditions. The *Ae. aegypti* expresses long wavelength rhodopsin Aaop1 on all R1–6 photoreceptors and most R8 photoreceptors. These photoreceptors alter the cellular location of Aaop1 and rearrange their photosensitive rabdomeric membranes daily [61].

The effects of temperature on *Ae. aegypti* larval behavior have been examined for different food availability conditions at breeding sites. With reduced food availability and low temperature, the larvae spend more time and energy to feed. At the temperature extremes, there are negative impacts on feeding behavior and larval activity [62].

Increasing viscosity by elevating the concentration of dissolved solids imposes unfavorable conditions, including limiting movement of the organisms. However, females adapt, choosing to perform oviposition even under unfavorable conditions. This phenomenon shows that for *Ae. aegypti*, the water physicochemical quality is not the most important parameter, but rather the conditions that the breeding site presents for the development and survival of the immature individuals are crucial.

Albeny-Simões et al. [63] showed that the choice of breeding grounds for oviposition and reproduction with aquatic larvae is influenced by the risk of offspring mortality and survival, and food availability. Studies revealed that *Ae. aegypti* is attracted to do the oviposition in breeding sites with predatory larvae (*Toxorhynchites* (Theobald) (Diptera: Culicidae)), as the density of dead larvae and metabolic products increase the bacterial abundance in the breeding location. Bacterial biomass and species composition are environmental determinants for the occurrence and abundance of *Aedes* spp., breeding sites with competing species [64, 65].

Soares Pinheiro et al. [66] evaluated *Ae. aegypti* viability in the Amazon region that has high humidity. The *Ae. aegypti* eggs have been stored in plastic cups, paper envelopes or plastic bags and maintained in internal and external areas for different times. Overall, the storage form is important for egg viability. The authors concluded that *Ae. aegypti* viability in the Amazon region is maintained at high levels up to 4 months, at which time there are drastic reductions and hatchings up to 8 months occur at very low percentages. These results show that excess humidity and temperature can induce constant stimuli to hatching eggs and possibly due to water stress, reduce their viability. Comparatively, in dry environments this viability can be preserved for up to 492 days [59]. In this regard, the selection of suitable place for oviposition is fundamental for the distribution and establishment of *Ae. aegypti*. Studies indicate that *Ae. aegypti* females are not attracted to breeding grounds with clean water. This flexibility to choose breeding sites for posture, despite the apparent risks to reproduction, is an important fact. This posture behavior demonstrates the adaptability of the insect to the different environmental situations, even though they appear to represent unfavorable conditions [61, 62].

Additionally, breeding sites of principal dengue virus spreader *Ae. aegypti* [67] and secondary vector *Ae. albopictus* [68] or both [69] can be stopped through biological control [70]; with environmental modifications [71]; utilizing copepods (H. Milne-Edwards) (Copepoda: Cyclopidae) [72]; and using ecological services by frogs, toads and tadpoles (Anura) [73].

3. Conclusions

Studies aimed the understanding of the life cycle and development of Culicidae mosquitoes and their association with breeding sites are extremely important, since the main control actions are performed in these microhabitats. Although many researchers have devoted themselves to this topic, elucidating the main factors associated with the development of immature forms and seeking new control alternatives, what we have observed is that these insects can colonize new microhabitats, expanding to new areas and reaching high levels of infestation, resulting in outbreaks and epidemics of some arboviruses, such as dengue, chikungunya and zika.

The main factor contributing to this fact is the high adaptive power that these insects have in colonizing new breeding sites, the natural ones, but especially the artificial ones, located in urban environments, which are increasingly available due to the lack of investment in the improvement of sanitary conditions in cities, disorderly expansion of urban centers, increase of consumable products in disposable packaging, among others.

The available studies on the type, frequency and colonization of *Aedes* spp. breeding sites show that artificial breeding sites are preferred, and can be colonized by *Aedes* and other associated species. Natural breeding sites in plants, rocks and organic remains can be colonized for long periods, preferably by *Ae. albopictus* and occasionally by *Ae. aegypti*. These preferences can be reversed based on climatic influences, food availability, and the breeding water physicochemical and biological conditions. The breeding sites most suitable for *Aedes* spp. reproduction have a considerable amount of decomposing organic matter and conditions that guarantee a reproductive advantage to these species.

Mosquito population control is dependent on adherence and awareness of populations and knowledge of breeding conditions. Population control interventions with new products with potentially wide use or efficient strategies that are based on habits and ecological knowledge of breeding sites can be advantageous for *Ae. aegypti* population and reproduction control.

Acknowledgements

Acknowledgments are due to CNPq (477012/2013-2\307390/2016-0) and FUNDECT (35308.522.6490.25062016) for the financial supports.

DOI: http://dx.doi.org/10.5772/intechopen.88852

Author details

Márcia Ramos Jorge[1], Antonio Pancrácio de Souza[2], Ricardo Augusto dos Passos[3,4], Silvia Maria Martelli[5], Chaiane Regina Rech[1], Alexeia Barufatti[6], Bruno do Amaral Crispim[6], Helina dos Santos Nascimento[6] and Eduardo José de Arruda[1]*

1 Faculty of Exact Sciences and Technology, Federal University of Grande Dourados, Dourados, Brazil

2 Institute of Biosciences – InBio, Federal University of Mato Grosso do Sul, Campo Grande, Brazil

3 Postgraduate Program in Entomology and Biodiversity Conservation, Faculty of Biological and Environmental Sciences, Federal University of Grande Dourados UFGD, Mato Grosso do Sul, Brazil

4 Laboratory of Physiology and Control of Arthropod Vectors – Oswaldo Cruz Foundation, Rio de Janeiro, Brazil

5 Faculty of Engineering, Federal University of Grande Dourados, Dourados, Brazil

6 Faculty of Biological and Environmental Sciences, Federal University of Grande Dourados, Dourados, Brazil

*Address all correspondence to: ejarruda@gmail.com

References

[1] Donalísio MR, Glasser CM. Vigilância entomológica e controle de vetores do dengue. Revista Brasileira de Epidemiologia. 2002;**5**(3):259-279. DOI: 10.1590/S1415-790X2002000300005

[2] Kotsakiozi P, Evans BR, Gloria-Soria A, Kamgang B, Mayanja M, Lutwama J, et al. Population structure of a vector of human diseases: *Aedes aegypti* in its ancestral range. Africa Ecology and Evolution. 2018;**8**(16):7835-7848. DOI: 10.1002/ece3.4278

[3] World Health Organization. Emergencies Preparedness, Response: Yellow Fever – Brazil [Internet]. 2017. Available from: http://www.who.int/csr/don/20-march-2017-yellow-fever-brazil/en/ [Accessed: 06 June 2019]

[4] Wong J, Stoddard ST, Astete H, Morrison AC, Scott TW. Oviposition site selection by the dengue vector *Aedes aegypti* and its implications for dengue control. PLoS Neglected Tropical Diseases. 2011;**5**(4):e1015. DOI: 10.1371/journal.pntd.0001015

[5] de Castro FP Jr, Martins WFS, de Lucena Filho ML, de Almeida RP, Beserra EB. Ciclos de vida comparados de *Aedes aegypti* (Diptera, Culicidae). do semiárido da Paraíba. 2013;**103**(2):118-123. DOI: 10.1590/S0073-47212013000200006

[6] Rozeboom LE, Rosen L, Ikeda J. Observations on oviposition by *Aedes* (S.) *albopictus* Skuse and *A.*(S.) *polynesiensis* Marks in nature. Journal of Medical Entomology. 1973;**10**(4):397-399. DOI: 10.1093/jmedent/10.4.397

[7] Chadee DD. Oviposition strategies adopted by gravid *Aedes aegypti* (L.) (Diptera: Culicidae) as detected by ovitraps in Trinidad, West Indies (2002-2006). Acta Tropica. 2009;**111**(3):279-283. DOI: 10.1016/j.actatropica.2009.05.012

[8] Reiter P. Oviposition, dispersal, and survival in *Aedes aegypti*: Implications for the efficacy of control strategies. Vector Borne and Zoonotic Diseases. 2007;**7**(2):261-273. DOI: 10.1089/vbz.2006.0630

[9] Barbosa AA, de Silva MAN. *Aedes* (Stegomyia) *albopictus* (Skuse)(Diptera, Culicidae), preference for oviposition site related with homospecific immatures presence, under laboratory conditions. Revista Brasileira de Zoologia. 2002;**19**(4):1147-1152. DOI: 10.1590/S0101-81752002000400020

[10] Madeira N, Macharelli C, Carvalho L. Variation of the oviposition preferences of *Aedes aegypti* in function of substratum and humidity. Memórias do Instituto Oswaldo Cruz. 2002;**97**(3):415-420. DOI: 10.1590/S0074-02762002000300025

[11] Mitchell-Foster K, Ma B, Warsame-Ali S, Logan C, Rau M, Lowenberger C. The influence of larval density, food stress, and parasitism on the bionomics of the dengue vector *Aedes aegypti* (Diptera: Culicidae): Implications for integrated vector management. Journal of Vector Ecology. 2012;**37**(1):221-229. DOI: 10.1111/j.1948-7134.2012.00220.x

[12] Rey JR, O'Connell SM. Oviposition by *Aedes aegypti* and *Aedes albopictus*: Influence of congeners and of oviposition site characteristics. Journal of Vector Ecology. 2014;**39**(1):190-196. DOI: 10.1111/j.1948-7134.2014.12086.x

[13] Instituto Oswaldo Cruz (IOC/Fiocruz). *Aedes aegypti*: introdução aos aspectos científicos do vetor. Modulo 3: hábitos e criadouros [Internet]. 2011. Available from: http://auladengue.ioc.fiocruz.br/?p=78 [Accessed: 15 May 2019]

[14] de Silva IG, Camargo MF, Elias M, Elias CN. Ciclo evolutivo de *Aedes*

(Stegomyia) *aegypti* (Linnaeus, 1762) (Diptera, Culicidae). Revista de Patologia Tropical. 1993;**22**(1). DOI: 10.5216/rpt.v22i1.20069

[15] Liu-Helmersson J, Stenlund H, Wilder-Smith A, Rocklöv J. Vectorial capacity of *Aedes aegypti*: Effects of temperature and implications for global dengue epidemic potential. PLoS One. 2014;**9**(3):e89783. DOI: 10.1371/journal.pone.0089783

[16] Ote M, Kanuka H. A highly secure method for rearing *Aedes aegypti* mosquitoes. Tropical Medicine and Health. 2018;**46**(16). DOI: 10.1186/s41182-018-0098-5

[17] Getachew D, Tekie H, Gebre-Michael T, Balkew M, Mesfin A. Breeding sites of *Aedes aegypti*: potential dengue vectors in Dire Dawa, East Ethiopia. Interdisciplinary perspectives on infectious diseases. 2015;**2015**:8. DOI: 10.1155/2015/706276

[18] Arduino MB, de Ávila GDO. Aspectos físico-químicos da água de criadouros de *Aedes aegypti* em ambiente urbano e as implicações para o controle da dengue. Revista de Patologia Tropical. 2015;**44**(1):89-100. DOI: 10.5216/rpt.v44i1.34801

[19] World Health Organization. Dengue Control [Internet]. 2014. Available from: http://www.who.int/denguecontrol/en/index.html [Accessed: 15 May 2019]

[20] Abe M, McCall PJ, Lenhart A, Villegas E, Kroeger A. The Buen pastor cemetery in Trujillo, Venezuela: Measuring dengue vector output from a public area. Tropical Medicine and International Health. 2005;**10**(6):597-603. DOI: 10.1111/j.1365-3156.2005.01428.x

[21] Surendran SN, Jude PJ, Thabothiny V, Raveendran S, Ramasamy R. Pre-imaginal development of *Aedes aegypti* in brackish and fresh water urban domestic wells in Sri Lanka. Journal of Vector Ecology. 2012;**37**(2):471-474. DOI: 10.1111/j.1948-7134.2012.00254.x

[22] Richardson BA. The bromeliad microcosm and the assessment of faunal diversity in a neotropical forest. Biotropica. 1999;**31**(2):321-336. DOI: 10.1111/j.1744-7429.1999.tb00144.x

[23] Barrera R, Amador M, Diaz A, Smith J, Munoz-Jordan J, Rosario Y. Unusual productivity of *Aedes aegypti* in septic tanks and its implications for dengue control. Medical and Veterinary Entomology. 2008;**22**(1):62-69. DOI: 10.1111/j.1365-2915.2008.00720.x

[24] Arduino MB, Marques GRAM, Serpa LLN. Registro de larvas e pupas de *Aedes aegypti* e *Aedes albopictus* em recipientes com água salina em condições naturais. BEPA Boletim Epidemiológico Paulista (Online). 2010;**7**(83):22-28

[25] Manrique-Saide P, Arisqueta-Chablé C, Geded-Moreno E, Herrera-Bojórquez J, Uc V, Chablé-Santos J, et al. An assessment of the importance of subsurface catch basins for *Aedes aegypti* adult production during the dry season in a neighborhood of Merida, Mexico. Journal of the American Mosquito Control Association. 2013;**29**(2): 164-168. DOI: 10.2987/12-6320R.1

[26] Beserra EB, Fernandes CRM, de Sousa JT, de Freitas EM, Santos KD. Efeito da qualidade da água no ciclo de vida e na atração para oviposição de *Aedes aegypti* (L.) (Diptera: Culicidae). Neotropical Entomology. 2010;**39**(6):1016-1023. DOI: 10.1590/S1519-566X2010000600026

[27] Fader JE, Juliano SA. Oviposition habitat selection by container-dwelling mosquitoes: Responses to cues of larval and detritus abundances in the field. Ecological Entomology.

2014;**39**(2):245-252. DOI: 10.1111/een.12095

[28] de Silva VC, Scherer PO, Falcão SS, Alencar J, Cunha SP, Rodrigues IM, et al. Diversidade de criadouros e tipos de imóveis freqüentados por *Aedes albopictus* e *Aedes aegypti*. Revista de Saúde Pública. 2006;**40**(6):1106-1111. DOI: 10.1590/S0034-89102006000700021

[29] Gil LHS, Katsuragawa TH, Lima AADE, Tada MS, Ozaki LS, Julião GR. Rudimentary cesspits as breeding sites for *Aedes aegypti* in urban areas of northern Brazil. Revista Pan-Amazônica de Saúde. 2015;**6**(3):73-80. DOI:10.5123/S2176-62232015000300010

[30] Rao BB, Harikumar PS, Jayakrishnan T, George B. Characteristics of *Aedes* (Stegomyia) *albopictus* Skuse (Diptera: Culicidae) breeding sites. Southeast Asian Journal of Tropical Medicine and Public Health. 2011;**42**(5):1077-1082

[31] Olayemi IK, Omalu ICJ, Famotele OI, Shegna SP, Idris B. Distribution of mosquito larvae in relation to physico-chemical characteristics of breeding habitats in Minna, North Central Nigeria. Reviews in Infection. 2010;**1**(1):49-53

[32] Fillinger U, Sombroek H, Majambere S, van Loon E, Takken W, Lindsay SW. Identifying the most productive breeding sites for malaria mosquitoes in The Gambia. Malaria Journal. 2009;**8**(1):62. DOI: 10.1186/1475-2875-8-62

[33] Clark TM, Flis BJ, Remold SK. pH tolerances and regulatory abilities of freshwater and euryhaline Aedine mosquito larvae. Journal of Experimental Biology. 2004;**207**: 2297-2304. DOI: 10.1242/jeb.01021

[34] Navarro DMAF, de Oliveira PES, Potting RPJ, Brito AC, Fital SJF, Sant'Ana AEG. The potential attractant or repellent effects of different water types on oviposition in *Aedes aegypti* L. (Dipt., Culicidae). Journal of Applied Entomology. 2003;**127**(1):46-50. DOI: 10.1046/j.1439-0418.2003.00690.x

[35] Gopalakrishnan R, Das M, Baruah I, Veer V, Dutta P. Physicochemical characteristics of habitats in relation to the density of container-breeding mosquitoes in Asom, India. Journal of Vector Borne Diseases. 2013;**50**(3):215

[36] Umar A, Don Pedro K. The effects of pH on the larvae of Ae. aegypti and Cx. quinquefasciatus. International Journal of Pure and Applied Sciences. 2008;**2**:58-62

[37] Tun-Lin W, Kay BH, Barnes A. Understanding productivity, a key to *Aedes aegypti* surveillance. The American Journal of Tropical Medicine and Hygiene. 1995;**53**(6):595-601. DOI: 10.4269/ajtmh.1995.53.595

[38] Ilkenhans T, Poulston S, Rowsell L, Smith AWJ, Terry LA. Development of a new palladium-based ethylene scavenger. In: Ramina A, Chang C, Giovannoni J, Klee H, Perata P, Woltering E, editors. Advances in Plant Ethylene Research. Dordrecht: Springer; 2007. pp. 211-213. DOI: 10.1007/978-1-4020-6014-4_44

[39] Ricklefs R. A Economia da Natureza. 5th ed. Rio de Janeiro: Ed. Guanabara Koogan; 2001 503p

[40] Superintendência de Controle de Endemias. Normas e Recomendações Técnicas para Vigilância e Controle do *Aedes aegypti* no Estado de São Paulo; 2002. 69p

[41] Laird M. The Natural History of Larval Mosquito Habitats. London: Academic Press Ltd; 1988. 555p

[42] Dom NC, Ahmad AH, Ishak AR, Ismail R. Assessing the risk of dengue

fever based on the epidemiological, environmental and entomological variables. Procedia - Social and Behavioral Sciences. 2013;**105**:183-194. DOI: 10.1016/j.sbspro.2013.11.019

[43] Nazri CD, Hashim A, Rodziah I, Hassan A, Yazid AA. Utilization of geoinformation tools for dengue control management strategy: A case study in Seberang Prai, Penang Malaysia. International Journal of Remote Sensing Applications. 2013;**3**(1):11-17

[44] Thangamathi P, Ananth S, Kala N, Maheshwari R, Gnanasoundrai A, Nagamani N. Seasonal variations and physicochemical characteristics of the habitats in relation to the density of dengue vector *Aedes aegypti* in Thanjavur, Tamil Nadu, India. International Journal of Science and Nature. 2014;**5**(2):271-276

[45] Zeidler JD, Acosta POA, Barreto PP, Cordeiro JS. Vírus dengue em larvas de *Aedes aegypti* e sua dinâmica de infestação, Roraima. Brasil Revista de Saúde Pública. 2008;**42**:986-991. DOI: 10.1590/S0034-89102008005000055

[46] Hasselschwert D, Rockett CL. Bacteria as ovipositional attractants for *Aedes aegypti* (Diptera: Culicidae). The Great Lakes Entomologist. 1988;**21**(4):163-168

[47] Trexler JD, Apperson CS, Zurek L, Gemeno C, Schal C, Kaufman M, et al. Role of bacteria in mediating the oviposition responses of *Aedes albopictus* (Diptera: Culicidae). Journal of Medical Entomology. 2003;**40**(6):841-848. DOI: 10.1603/0022-2585-40.6.841

[48] Poonam S, Paily KP, Balaraman K. Oviposition attractancy of bacterial culture filtrates: Response of *Culex quinquefasciatus*. Memórias do Instituto Oswaldo Cruz. 2002;**97**(3):359-362. DOI: 10.1590/S0074-02762002000300015

[49] Huang J, Miller JR, Chen S-C, Vulule JM, Walker ED. *Anopheles gambiae* (Diptera: Culicidae) oviposition in response to agarose media and cultured bacterial volatiles. Journal of Medical Entomology. 2006;**43**(3):498-504. DOI: 10.1093/jmedent/43.3.498

[50] Ponnusamy L, Schal C, Wesson DM, Arellano C, Apperson CS. Oviposition responses of *Aedes* mosquitoes to bacterial isolates from attractive bamboo infusions. Parasites & Vectors. 2015;**8**(1):486. DOI: 10.1186/s13071-015-1068-y

[51] Ponnusamy L, Xu N, Nojima S, Wesson DM, Schal C, Apperson CS. Identification of bacteria and bacteria-associated chemical cues that mediate oviposition site preferences by *Aedes aegypti*. Proceedings of the National Academy of Sciences. 2008;**105**(27):9262-9267. DOI: 10.1073/pnas.0802505105

[52] Ponnusamy L, Wesson DM, Arellano C, Schal C, Apperson CS. Species composition of bacterial communities influences attraction of mosquitoes to experimental plant infusions. Microbial Ecology. 2010;**59**(1):158-173. DOI: 10.1007/s00248-009-9565-1

[53] Seenivasagan T, Guha L, Parashar BD, Agrawal OP, Sukumaran D. Olfaction in Asian tiger mosquito *Aedes albopictus*: Flight orientation response to certain saturated carboxylic acids in human skin emanations. Parasitology Research. 2014;**113**(5):1927-1932. DOI: 10.1007/s00436-014-3840-x

[54] Thorn RMS, Reynolds DM, Greenman J. Multivariate analysis of bacterial volatile compound profiles for discrimination between selected species and strains in vitro. Journal of Microbiological Methods. 2011;**84**(2):258-264. DOI: 10.1016/j.mimet.2010.12.001

[55] Medeiros PFV. Investigações sobre os mecanismos de resistência em larvas e adultos de *Aedes aegypti*, Linnaeus, 1762 [thesis]. Rio de Janeiro: Fundação Osvaldo Cruz/Instituto Osvaldo Cruz-Fiocruz; 2011

[56] Guirado MM, Bicudo HEMC. Alguns aspectos do controle populacional e da resistência a inseticidas em *Aedes aegypti* (Diptera, Culicidae). Boletim Epidemiológico Paulista. 2009;**6**(64):5-14

[57] Mondini A, Chiaravalloti-Neto F. Spatial correlation of incidence of dengue with socioeconomic, demographic and environmental variables in a Brazilian city. Science of the Total Environment. 2008;**393**(2-3):241-248. DOI: 10.1016/j.scitotenv.2008.01.010

[58] Antonio FJ, Itami AS, Picoli SD, Teixeira JJV, de Santos MR. Spatial patterns of dengue cases in Brazil. PLoS One. 2017;**12**(7):e0180715. DOI: 10.1371/journal.pone.0180715

[59] Silva HHG, Silva IG. Influência do período de quiescência dos ovos sobre o ciclo de vida de *Aedes aegypti* (Linnaeus, 1762) (Diptera, Culicidae) em condições de laboratório. Revista da Sociedade Brasileira de Medicina Tropical. 1999;**32**(4):349-355. DOI: 10.1590/S0037-86821999000400003

[60] Beserra EB, de Castro Jr. FP. Biologia comparada de populações de *Aedes* (Stegomyia) *aegypti* (L.)(Diptera: Culicidae) da Paraíba. Neotropical Entomology. 2008;**37**(1):81-85. DOI: 10.1590/S1519-566X2008000100012

[61] Hu X, Leming MT, Metoxen AJ, Whaley MA, O'Tousa JE. Light-mediated control of rhodopsin movement in mosquito photoreceptors. Journal of Neuroscience. 2012;**32**(40):13661-13667. DOI: 10.1523/JNEUROSCI.1816-12.2012

[62] Reiskind MHJ, Shawn M. Late-instar behavior of *Aedes aegypti* (Diptera: Culicidae) larvae in different thermal and nutritive environments. Journal of Medical Entomology. 2015;**52**(5):789-796. DOI: 10.1093/jme/tjv088

[63] Albeny-Simões D, Murrell EG, Elliot SL, Andrade MR, Lima E, Juliano SA, et al. Attracted to the enemy: *Aedes aegypti* prefers oviposition sites with predator-killed conspecifics. Oecologia. 2014;**175**(2):481-492. DOI: 10.1007/s00442-014-2910-1

[64] Murrell EG, Damal K, Lounibos LP, Juliano SA. Distributions of competing container mosquitoes depend on detritus types, nutrient ratios, and food availability. Annals of the Entomological Society of America. 2011;**104**(4):688-698. DOI: 10.1603/AN10158

[65] Afify A, Galizia CG. Chemosensory cues for mosquito oviposition site selection. Journal of Medical Entomology. 2015;**52**(2):120-130. DOI: 10.1093/jme/tju024

[66] Soares-Pinheiro VC, Dasso-Pinheiro W, Trindade-Bezerra JM, Tadei WP. Eggs viability of *Aedes aegypti* Linnaeus (Diptera, Culicidae) under different environmental and storage conditions in Manaus, Amazonas, Brazil. Brazilian Journal of Biology. 2017;**77**(2):396-401. DOI: 10.1590/1519-6984.19815

[67] Sarwar M. Proposals for the control of principal dengue fever virus transmitter *Aedes aegypti* (Linnaeus) mosquito (Diptera: Culicidae). Journal of Ecology and Environmental Sciences. 2014;**2**(2):24-28

[68] Sarwar M. Role of secondary dengue vector mosquito *Aedes albopictus* Skuse (Diptera: Culicidae) for dengue virus transmission and its coping. International Journal of Animal Biology. 2015;**1**(5):219-224

[69] Sarwar M. Proposing solutions for the control of dengue fever virus carrying mosquitoes (Diptera: Culicidae) *Aedes aegypti* (Linnaeus) and *Aedes albopictus* (Skuse). Journal of Pharmacology and Toxicological Studies. 2014;**2**(1):1-6

[70] Sarwar M. Reducing dengue fever through biological control of disease carrier *Aedes* mosquitoes (Diptera: Culicidae). International Journal of Preventive Medicine Research. 2015;**1**(3):161-166

[71] Sarwar M. Stopping breeding of dengue virus spreader *Aedes* mosquitoes (Diptera: Culicidae) with environmental modifications. International Journal of Bioinformatics and Biomedical Engineering. 2015;**1**(2):169-174

[72] Sarwar M. Elimination of dengue by control of *Aedes* vector mosquitoes (Diptera: Culicidae) utilizing copepods (Copepoda: Cyclopidae). International Journal of Bioinformatics and Biomedical Engineering. 2015;**1**(1):53-58

[73] Sarwar M. Controlling dengue spreading *Aedes* mosquitoes (Diptera: Culicidae) using ecological services by frogs, toads and tadpoles (Anura) as predators. American Journal of Clinical Neurology and Neurosurgery. 2015;**1**(1):18-24.7

Chapter 5

Characterization of the Proliferation Sites of *Aedes aegypti* (Diptera: Culicidae) in the Artificial Breeding Sites of Caxias, Maranhão, Brazil

Aylane Tamara dos Santos Andrade,
Juliana Maria Trindade Bezerra
and Valéria Cristina Soares Pinheiro

Abstract

The *Aedes aegypti* (Diptera: Culicidae) is dispersed throughout the Brazilian territory, being a transmitter of dengue and other arboviruses. This study analyzes the main breeding of *A. aegypti* in semi-arid tropical region. Samples have been collected for 12 months in the dry and rainy periods to watch the main breeding and characteristics of the proliferation sites. Most of the positive containers have been from the storage group (75.96%) found with immature forms mainly in the rainy season, with a predominance of containers having a height superior to 50 cm, protected from the sun and with organic matter. Theses breeding sites, where *A. aegypti* performs oviposition and survives in the dry season too.

Keywords: mosquitoes, oviposition, control, tropical climate, semi-arid, vector

1. Introduction

Aedes aegypti (Linnaeus) and *Aedes albopictus* (Skuse) are mosquitoes of the genus *Aedes*, belonging to the family Culicidae, which stand out as the two main species of vectors transmitting viral diseases, such as dengue, which has four serotypes (DENV 1–4), viruses Chikungunya, Zika virus, among other arboviruses [1–4].

Dengue is currently the most prevalent arboviruses in the world. According to the World Health Organization, about half of the world's population lives at risk of infection [5]. The *A. aegypti* is incriminated by the transmission of dengue virus and has been circulating in several continents such as Asia, Africa, the Americas and Europe [6].

In Brazil, *A. aegypti* is the most important vector of dengue virus. Since the twentieth century, campaigns to combat this vector have been intensified; however, they have not been successful. The *A. aegypti* continues to disperse in various parts of the world where it transmits several arboviruses. This fact can also be seen throughout the Brazilian territory [6, 7].

The *A. aegypti* is a diurnal, anthropophilic mosquito with development of holometable, passing through the egg phase, four larval stages (L1, L2, L3 and L4), pupa and adult forms. The increase in the availability of human-made containers associated with climatic factors such as temperature rise, and variations in rainfall and relative humidity favor the vector development [8–11].

Among the control methods, the most important are the chemicals, which have been used since the 1940s with the discovery of synthetic insecticides. Although several studies prove the resistance of mosquito populations to different classes of insecticides, they are still widely used for the control of *Aedes*. We also have the mechanical control that aims at the removal of breeding grounds, and the biological control that uses living organisms considered ecologically safe for the environment, as an attempt to reduce the population density of the vector [12, 13].

In order to combat *A. aegypti*, it is essential to regularly implement public water supply systems, basic sanitation and the adequate packaging of municipal solid waste, because a large part of the population is not aware of the problems generated by poor packaging of household waste and containers [14, 15].

The present work aims to characterize the breeding sites most frequented by *A. aegypti*, to analyze the physical conditions in which these breeding sites are found to contribute for the actions to combat vector proliferation. As a consequence, outcome of this work will promote improvement in health and the environment, avoiding the disposal of waste in inappropriate places and an adequate packaging of domestic containers.

2. Materials and methods

2.1 Characterization of the study area

The study has been conducted in the city of Caxias - MA, located between the coordinates 04°51′32 "south latitude and 43°21'22" longitude west. Its estimated population in 2018 is 164,224 inhabitants [16]. The climate is hot and humid with an average temperature of 28.2°C, having averages of the maximum of 27°C and averages of the minimum of 17°C, with two defined seasons, like rainy (January to June) and dry (July to December). The pluviometric index is between 1100 and 1500 mm/year, being the Itapecuru basin the main municipality, since it occupies 3567.40 km^2 (68.4%) of the territory, in which there is a diversity of biomes and vegetation formation inserted in an ecotone area [17].

The area has been chosen randomly, where five zones have been selected according to the listing of districts of Caxias - MA, with records of dengue cases in the last 5 years.

2.2 Immature collections

2.2.1 Sampling

The city of Caxias presents a total of 55,769 properties. The zones drawn have a total of 6973 real estate [18]. They have been inspected 80 properties by neighborhood, chosen of random form. This sampling has been defined according to the formula for the Calculation of the Minimum Sample Size, proposed by [19] tolerating the margin of error of 5%.

The surveyed neighborhoods are among the most populous, have poor sanitation conditions, the sewers are open, some streets without adequate pavement, there is a regular lack of water from the public supply. Most homes have large backyards with trees and the presence of domestic animals such as chickens, cats, dogs and even pigs. It is common to find accumulated garbage and water containers that serve as drinking fountains for the animals.

The collections have been carried out during a year with a survey of 400 residences in the rainy season with repetition of the same ones in the dry season. All artificial containers that showed potential for *A. aegypti* proliferation have been analyzed, following the methodology of the National Health Foundation with inspection of the internal and external residences [20, 21].

The vessels inspected have been classified into groups according to following [22–25].

TR1 – vase (plant pots, and pot dishes);
TR2 – bottle (bottles, cans, plastic);
TR3 – tire (all types of tire);
TR4 – building materials and car parts (materials used for construction and parts of cars);
TR5 – storage (tank, barrel, filter, pot, bucket, drum);
TR6 – fixed (drains, grease box and wells);
TR7 – others (all containers that do not fit the others).

The positive containers, which have been found in the presence of immature classified as to height (separated into three categories, 0–50, 51–100 and 100 cm above), exposure of breeding (total or partial exposure to the sun and/or shadow), and the presence or absence of organic matter.

In the containers with larvae/pupae all specimens have been collected. In the deposits where it is not possible the total withdrawal of the immature ones was made in a numerical calculation of the area multiplied by the volume of the breeding place that measured with the aid of millimeter vial and tape measure. This calculation is used by endemic agents to estimate the population density of large breeding sites [21].

The larvae and pupae have been captured using plastic pipettes, and collected and conditioned in 10 ml hemolysis tubes containing 70% alcohol. Each tube stipulated a maximum number of 20 specimens, and the flasks have been identified with label containing the location data and type of deposit where they collected.

Subsequently, they have been sent to the Laboratory of Medical Entomology (LABEM) to identify the specimens, and quantify and record the results in a bulletin for evaluation of larval density.

The identification has been made through the direct observation of the morphological characters evidenced to the stereoscopic and the light microscope using the dichotomous key proposed [8, 9].

The number of immatures collected in the five zones of the municipality of Caxias, Maranhão, has been evaluated for the normal distribution, using the Shapiro-Wilk test. As these do not fit to the normal distribution, corresponding non-parametric analyzes have been used.

The Kruskal-Wallis test (H) has been used to verify the differences between the number of immatures between zones, months and collection containers. The value of H calculated for each analysis has been compared to the value of H defined in the quantis table for Kruskal-Wallis test statistics. Whenever the calculated value is greater than that presented in the table at a certain degree of freedom, considered the number of groups compared, and where the value of p is less than or equal to 0.05, the hypothesis of medium males used, and continue with the Dunn test comparison of medians posteriori [26].

All analyzes have been performed using the Statistical Package for Social Sciences version 20.0 (SPSS Inc., Chicago, IL), the information compiled into database, mounted on Excel spreadsheets (2013) and Graph Pad Prism version 5.03 (La Jolla CA, USA), and from these obtained the percentages for the elaboration of tables and figures.

The values of p, significant at the mentioned level (5%), have been highlighted in the text, tables and graphs with an asterisk (*). For $p \leq 0.05$ an asterisk (*) has been used, for $p \leq 0.01$ two asterisks (**) have been used and for $p \leq 0.0001$ three asterisks (***) adopted (26).

3. Results

3.1 Collection cycles

When comparing the seven groups of containers surveyed, there is no significant difference between the number of containers found in the peri (H = 2399; p = 0.6628) and intradomicile (H = 1275; p = 0.8656) of residences. As well as there is no difference between the medians of vessels surveyed among the five zones throughout the study period (H = 2026; p = 0.7309).

The western zone has the highest percentage of containers in the storage group, with 637 (27.33%), of these 436 (68.44%) have been found in the peridomicile and 201 (31.55%) in intradomicile. It followed the Central zone with 530 (22.74%), North zone with 444 (19.05%) and South zone with 397 (17.03%) containers. The lowest percentage of inspected containers has been registered in the eastern zone with 322 (13.81%), of these 268 (83.23%), belongs to the peridomicile and 54 (16.77%) in the intradomicile.

When analyzing the presence of immatures in the positive containers of the seven groups, a difference is observed between the medians of *A. aegypti* in these groups (H = 30.74; p < 0.0001). According to the Dunn's test the storage group has predominant in relation to the vessels group (p < 0.01), construction material (p < 0.01), fixed (p < 0.01) and others (p < 0.01) (**Figure 1**).

The storage group has the largest number of containers surveyed in all study areas, with 2330 (55.17%) containers, of which 1653 (70.95%) have been found in the peridomicile and 677 (29.05%) in intradomiciliary. Followed by the bottle group with 1028 (24.34%), tires with 244 (5.77%) and others with 232 (5.49%). The lesser containers belong to the group pots 145 (3.43%), building material with 126 (2.98%) and fixed 118 (2.79%).

Positive containers exposed to the sun/shade showed a difference between the medians (H = 12.15, p = 0.0023). The containers have been positioned in the shade predominantly in relation to those in the shade (p < 0.01) – Dunn's test (**Figure 2**).

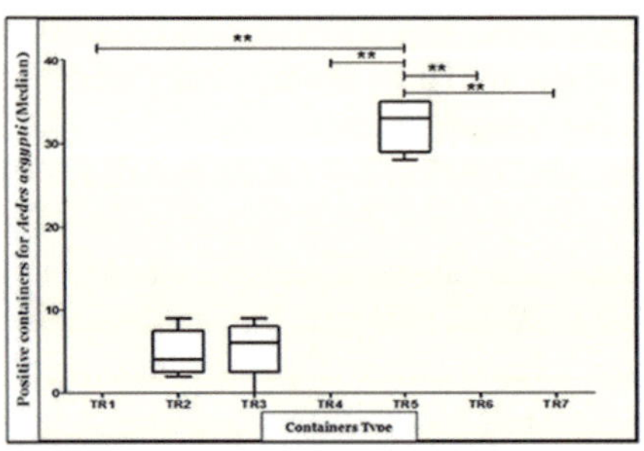

Figure 1.
Positive containers in the five collection zones in the municipality of Caxias-Maranhão, from January to December, 2016. Subtitle. TR1: Vessels, TR2: Bottles, TR3: Tires, TR4: Building Materials, TR5: Storage, Tr6: Fixed, TR7: Others.

When the exposure of the containers to the environment has been analyzed, it is observed in the rainy season predominance of immatures placed in containers in the sun with 1365 (75.74%) and shade with 437 (24.25%). In the dry season, the largest number of specimens is found in the shade with 602 (51.89%) and sun with 558 (48.10%). No immature is found in partially covered containers for the storage group (**Figure 3**).

When analyzing the positive containers with presence or absence of organic matter, a greater quantity of specimens in containers with organic substrates, have been registered in the rainy and dray seasons 1629 (90.55%) and 869 (74.91%), respectively. For containers without the presence of organic matter has been registered in the rainy season 170 (9.44%) immature and 291 (25.08%) in the dry season (**Figure 4**).

The containers with height between 51 and 100 cm recorded the highest number of *A. aegypti* in both rainy and dry seasons, with 1299 (72%) and 781 (67%), respectively. Following the containers with height between 0 and 50 cm, in the rainy season with 431 (23%) specimens and 238 (21%) in the dry season.

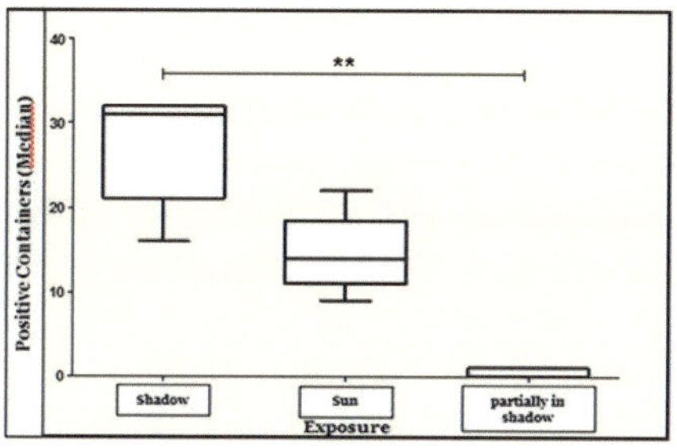

Figure 2.
Place of exposure of positive containers in the five collection zones from January to December 2016.

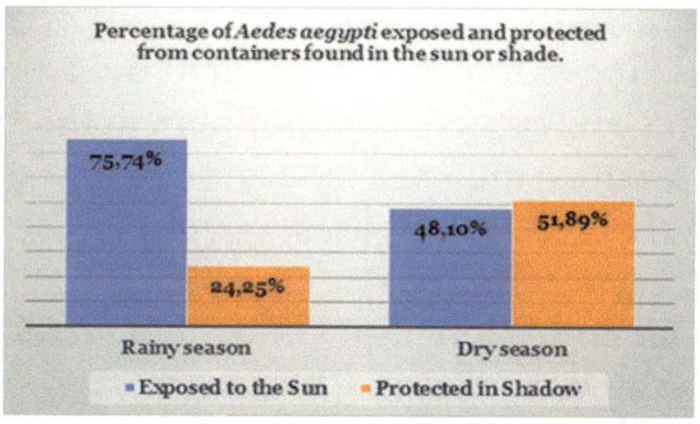

Figure 3.
Percentage of Aedes aegypti *found in containers exposed to the environment during the rainy and dry seasons, from January to December 2016.*

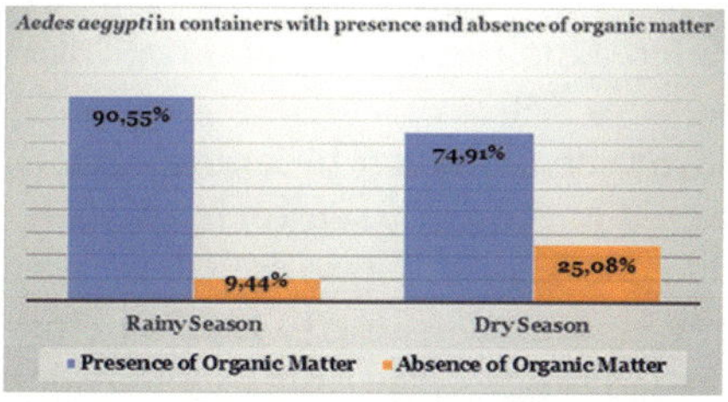

Figure 4.
Number of Aedes aegypti *found in containers with presence or absence of organic matter during the rainy and dry seasons, from January to December, 2016.*

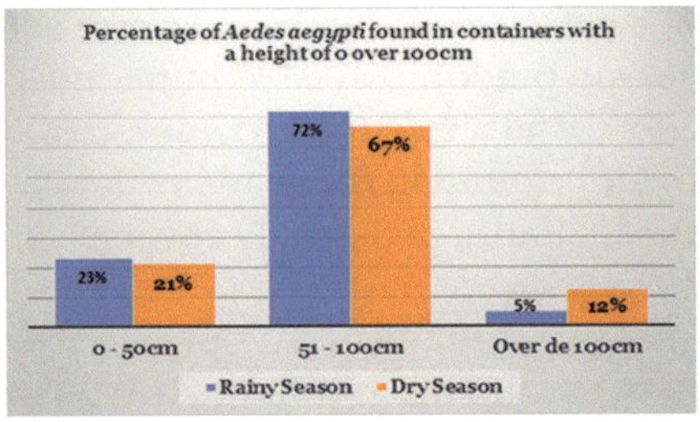

Figure 5.
Number of Aedes aegypti *distributed in containers with height between 0 and 100 cm, from January to December 2016.*

The containers above 100 cm registered 87 (5%) immature in the rainy season and 141 (12%) in the dry season (**Figure 5**).

4. Discussion

The data obtained in this research in relation to the main containers used by *A. aegypti* in urban area of tropical climate do not show statistical difference to the location in the peridomicile and the intradomicile of the residences, however, only the containers of the peridomicile have been found positive for immature forms, especially the storage group, which recorded the highest number of immature *A. aegypti*. Studies on the survey of potential *A. aegypti* breeding sites in the state of Ceará show that large containers such as water boxes hold *A. aegypti* larvae, and that containers found around the university such as cups, plastic bottles aluminum lunchboxes and small plastic containers contained *A. aegypti* eggs. In the present work we also found useless containers with foci of *A. aegypti* [27].

In the municipality of Caxias, Maranhão, other works have already been developed to characterize breeding sites most frequented by *A. aegypti* and *A. albopictus* [24, 25, 28] and also found preference for containers in the storage group, mainly tanks, barrels and basins, which are used by a large part of the local population to store water to be used in their daily activities, in view of the frequent lack of water in several districts of different municipal areas.

The routine of the residents to store water and other factors such as bad waste packaging and the local weather conditions, besides that dispersion of vector *A. aegypti* throughout the city, contribute to the occurrence of some diseases transmitted by *A. aegypti*, particularly Chikungunya and Zika, which is the target of an epidemic in this city during the accomplishment of this research (complementary information).

Artificial containers increasingly gain space in the urban environment, and inappropriate disposal favors conditions for vector mosquito breeding, among them *A. aegypti*, has demonstrated a high adaptive potential with the ability to colonize the most diverse breeding sites [29].

In relation to the exposure of *A. aegypti* breeding sites, the largest number of specimens has been collected in the rainy season in containers placed in the sun and in the dry season in shaded containers. These data demonstrate the adaptive evolution of *A. aegypti* that for decades according to [30], in Florida, [31] in the United States and by [32], in Brazil, has preferred only by shady containers, protected by trees, branches or slopes.

When analyzing the habitat of *A. aegypti*, it has been observed in a greater number of specimens in containers with organic matter presence, both in the rainy season and in the dry season [33]. When analyzing the effect of water quality on the life cycle of *A. aegypti,* there registered a significant decrease in the complete development of immature forms in relation to water and diet. This fact can justify the decrease in the number of specimens found in completely clean containers, without leaves, mosses, fruit residue or any other type of organic substrate.

Artificial containers are distinguished by their favorable conditions for proliferation of vector mosquitoes. The storage group is the main breeding site of *A. aegypti*, but other groups such as flasks and tires have also witnessed immature forms. It has been found that breeding sites between 0 and 50 cm deserve attention in the fight against *A. aegypti* because the accumulation of small containers such as bottles, glasses, sardine cans, soda cans and tires stacked in the peridomicile of homes or improperly discarded in the streets and corners contribute to the dispersion of this vector.

5. Conclusions

Most mosquitoes have a preference for the site of development, looking for conditions that help in their survival and proliferation. The residences offer ideal conditions for their growth and proliferation, since human blood that is part of their diet is easily found, as well as standing water and organic matter that enhances their growth.

The storage group was the most frequent breeding site for *A. aegypti*; these containers are often found in the municipality under study, and in rainy seasons deserve extra attention, as residents collect rainwater for use in domestic activities, significantly increasing the number of breeding sites in the municipality.

Disordering dump and inadequate maintenance of domestic containers mainly during periods of high rainfall, associated with favorable conditions of shade, height compatible with the ability to fly along with organic matter deposited in the bottom or edge of the breeding grounds favor the creation of these vectors. So, it becomes essential control measures and regular monitoring intra, peri and extra domicile residences, as well as information campaigns showing cases, notifications and mortality of diseases transmitted by *A. aegypti*.

The knowledge about the preferred sites for *A. aegypti* development is of fundamental importance as it contributes to the development of new strategies to control these mosquitoes.

Acknowledgements

Universidade Estadual do Maranhão (UEMA); Programa de Pós-Graduação em Biodiversidade, Ambiente e Saúde (PPGBAS); and Fundação de Pesquisa e Desenvolvimento Científico e Tecnológico do Maranhão (FAPEMA).

Author details

Aylane Tamara dos Santos Andrade[1], Juliana Maria Trindade Bezerra[2] and Valéria Cristina Soares Pinheiro[1*]

1 Universidade Estadual do Maranhão (UEMA), Caxias, Maranhão, Brazil

2 Universidade Federal de Minas Gerais (UFMG), Belo Horizonte, Minas Gerais, Brazil

*Address all correspondence to: vc_pinheiro@hotmail.com

References

[1] Miller BR, Ballinger ME. *Aedes albopictus* mosquitoes introduced into Brazil: Vector competence for yellow fever and dengue viruses. Transactions of the Royal Society of Tropical Medicine and Hygiene. 1988;**82**:476-477. DOI: 10.1016/0035-9203(88)90168-X

[2] Lourenço-De-Oliveira R, Vazeille M, Filippis AMB, Failloux AB. *Aedes aegypti* in Brazil: Genetically differentiated populations with high susceptibility to dengue and yellow fever viruses. Transactions of the Royal Society of Tropical Medicine and Hygiene. 2004;**98**:43-54. DOI: 10.1016/s0035-9203(03)00006-3

[3] Viennet E, Ritchie SA, Faddy HM, Williams CR, Harley D. Epidemiology of dengue in a high-income country: A case study in Queensland, Australia. Parasites & Vectors. 2014;7:379. DOI: 10.1186/1756-3305-7-379

[4] Souza LJ. Dengue, Zika e Chikungunya: Diagnosis, Treatment e Prevention. Rio de Janeiro: Rubio; 2016

[5] WHO. Dengue y dengue grave. Main factors [Internet]. 2019. Available from: https://www.who.int/news-room/fact-sheets/detail/dengue-and-severe-dengue [Accessed: 26 September 2019]

[6] Guzman MG, Halstead SB, Artsob H, Buchyp FJ, Gubler DJ. Dengue: A continuing global threat. Nature Reviews. Microbiology. 2010;**8**:S7-S16. DOI: 10.1038/nrmicro2460

[7] Epidemiological Bulletin. Department of Health Surveillance. Ministry of Health. Volume 50. April 2019. Monitoring of cases of urban arboviruses transmitted by Aedes (dengue, chikungunya e Zika) until Epidemiological Week of 12 de 2019 e Fast Index Survey for *Aedes aegypti* (LIRAa) [Internet]. 2019. Available from: http://portalarquivos2.saude.gov.br/images/pdf/2019/abril/30/2019-013-Monitoramento-dos-casos-de-arboviroses-urbanas-transmitidas-pelo-Aedes-publicacao.pdf [Accessed: 08 July 2019]

[8] Consoli R, Lourenço-De-Oliveira R. Main Mosquitoes of Sanitary Importance in Brazil. Rio de Janeiro: Fiocruz; 1994. p. 225

[9] Forattini OP. Medical Culicidology. Vol. 2. São Paulo: Publisher of the University of São Paulo; 2002

[10] Higa Y. Dengue vectors and their spatial distribution. Tropical Medicine and Health. 2011;**39**:S17-S27. DOI: 10.2149/tmh.2011-S04

[11] Viana DV, Ignotti E. The occurrence of dengue and meteorological variations in Brazil: Systematic review. Brazilian Journal of Epidemiology. 2013;**16**:240-256. DOI: 10.1590/S1415-790X2013000200002

[12] Ferreira LM, Silva-Filha MHL. Bacterial larvicides for vector control: Mode of action of toxins and implications for resistance. Biocontrol Science and Technology. 2013;**23**:1137-1168. DOI: 10.1080/09583157.2013.822472

[13] Zara ALDSA, Santos SMD, Fernandes-Oliveira ES, Carvalho RG, Coelho GE. Strategies for controlling *Aedes aegypti*: A review. Epidemiology and Health Services. 2016;**25**:391-404

[14] Oliveira GLAD. Prevention and Control of Dengue in the City of Sabará/MG: Analysis of Printed Educational Materials and Social Representations of Endemic Control Agents [thesis]. Belo Horizonte, MG, Brasil: Oswaldo Cruz Foundation, Research Center René Rachou; 2012

[15] Peres AC. Aedes: Expanding the focus. Communication and Health Journal Radis. National School of Public Health. ENSP Fiocruz; 2016. p. 161

[16] IBGE - Brazilian Institute of Geography and Statistics. Maranhão, Caxias, Infographics: General Data of the Municipality [Internet]. 2016. Available from: https://www.ibge.gov.br/ [Accessed: 01 October 2016]

[17] Souza IG. Invisible Cartography: Knowledge and Feelings of Caxias. Caxiense Academy of Letters: Caxias; 2015

[18] CCZ – Center for Zoonoses Control of Caxias-MA. Number of properties and districts of the municipality of Caxias/MA; 2017

[19] Barbetta PA. Statistics Applied to the Social Sciences. 5th ed. Florianópolis: UFSC; 2002. CDD: 300:21

[20] Brasil/FUNASA. National Foundation of Health. Dengue Instructions for Staff to Combat the Vector - Technical Standards Manual [Internet]. 2001. Available from: http://bvsms.saude.gov.br/bvs/publicacoes/funasa/man_dengue.pdf [Accessed: 21 June 2016]

[21] Brasil/SVS/MS. Ministry of Health. FUNASA – National Health Foundation. Dengue: Clinical and epidemiological characteristics. Epidemiological Surveillance Guide [Internet]. 2002. Available from: http://bvsms.saude.gov.br/bvs/publicacoes/dengue_aspecto_epidemiologicos_diagnostico_tratamento.pdf [Access: 21-06-2016]

[22] Pereira M. Productivity and larval habitats of *Aedes aegypti* in Santos, State of São Paulo [thesis]. Faculty of Public Health, University of São Paulo; 2001

[23] Pinheiro VCS, Tadei WP. Evaluation of the residual effect of temephos in *Aedes aegypti* larvae (Diptera: Culicidae) in artificial containers in Manaus, Amazonas, Brazil. Journal of Public Health. 2002;**18**:1529-1536

[24] Soares-Da-Silva J, Ibiapina SS, Bezerra JMT, Tadei WP, Pinheiro VCS. Variation in *Aedes aegypti* (Linnaeus) (Diptera, Culicidae) infestation in artificial containers in Caxias, state of Maranhão, Brazil. The Journal of the Brazilian Society of Tropical Medicine. 2012;**45**:174-179. DOI: 10.1590/S0037-86822012000200007

[25] Andrade ATS, Lobo KS, Soares-da-Silva J, Pinheiro VCS. Population density of *Aedes aegypti* (Linnaeus, 1762) and *Aedes albopictus* (Skuse, 1894) (Diptera: Culicidae) in artificial breeding of the city of Caxias, Maranhão. In: Fonseca RS, Barros MC, editors. Advances in Biological Sciences at the Center for Higher Studies of Caxias (CESC). Maranhão, São Luis: EDUEMA; 2018. pp. 135-152

[26] Siqueira AL, Tibúrcio JD. Statistics on health area – Conceitos, metodologias, aplicações e prática computacional. Coopmed: Belo Horizonte; 2011

[27] Neto TSC, Ramirez MTP, Galindo VR, Herculano LFS, Campello MVM. Levantamento de potenciais criadouros de *Aedes aegypti* no Campus do Itaperi da Universidade Estadual do Ceará. Medicina Veterinária. 2019;**13**:43-48. DOI: https://doi.org/10.26605/medvet-v13n1-2608

[28] Bezerra JMT, Santana INS, Miranda JP, Tadei WP, Pinheiro VCS. Breeding sites of Aedes aegypti (Linnaeus) (Diptera, Culicidae): A study about the containers diversity in dry and rainy seasons in a dengue-endemic city/ criadouros de Aedes aegypti (Linnaeus) (Diptera, Culicidae). Journal of Research in Health Sciences. 2018;**18**:102-107

[29] Diniz MTM, Medeiros JB. Mapeamento de focos de reprodução de aedes aegypti na cidade de caicó/rn com o auxílio de veículo aéreo não tripulado (mapping of breeding sites of *Aedes Aegypti* In Caicó/Rn City with use of unmanned aerial vehicle). Revista GeoNordeste. 2018;**2**:196-207

[30] Barrera R, Machado-Allison CE, Bulla LA. Breeders, larval density and segregation of niche in three urban culicidae. (Culex fatigans Wied., C. corniger Theo. y Aedes aegypti L.) in the cemetery of Caracas. Venezuelan Scientific Act. 1979;**30**:418-424

[31] Beier JC, Patricoski C, Travis M, Kranzfelder J. Influence of water chemical and environmental parameters on larval mosquito dynamics in tires. Environmental Entomology. 1983;**12**:434-438. DOI: 10.1093/ee/12.2.434

[32] Lopes J, Da Silva MA, Borsato AM, De Oliveira VD, Oliveira FJDA. *Aedes (Stegomyia) aegypti* L. and associated culicideofauna in an urban area of the south region, Brazil. Journal of Public Health. 1993;**27**:326-333. DOI: 10.1590/S0034-89101993000500002

[33] Beserra EB, Fernandes CR, Sousa JD, Freitas ED, Santos KD. Effect of water quality on the life cycle and on the attraction for oviposition of *Aedes aegypti* (L.) (Diptera: Culicidae). Neotropical Entomology. 2010;**39**:1016-1023

Chapter 6

Post-Embryonic Development of *Aedes* (*Stegomyia*) *aegypti* Linnaeus, 1762 at Different Temperatures and CO_2 Concentrations, and Their Influences on Hatching and Development of Stabilized Population

Ediane Oliveira do Amaral, André Felipe da Silva, William Ribeiro da Silva, Karina Rossi da Silva, Rosemary Aparecida Roque, Wanderli Pedro Tadei and João Antonio Cyrino Zequi

Abstract

This research aimed to verify biological parameters of *Aedes aegypti* Linnaeus, from Londrina, Paraná, in an incubator chamber (BOD) with different temperatures, and to analyze biological aspects of this mosquito from Manaus, Amazonas, in environments simulating the climatic conditions provided by the IPCC. In Londrina, the eggs were incubated for 10 days in BOD at different temperatures. The viability of eggs, number of adults, and mortality rate were analyzed later. In Manaus, the biological cycle time, number of adults, and mortality rate were analyzed in environmental rooms with different temperatures and CO_2 concentrations. The viability of eggs and the number of adults from Londrina was greater at 5 and 25°C, while the mortality rate of immatures was greater at 0°C; eggs incubated at 45°C did not hatch. Mosquitoes from Manaus completed the fastest biological cycle in room 4. The mortality percentage in the different instars for rooms 1, 2 and 4 was: 14.4; 28 and 53.6%, respectively. Thus, temperatures from 5 to 29.74°C were more appropriate since values outside these limits can cause deleterious effects on the species during its development, but the *A. aegypti* mosquitoes from Londrina and Manaus can benefit from the increase in temperature stipulated by the IPCC.

Keywords: arbovirus, mosquitoes, control, climate, quiescence

1. Introduction

Dengue is one of the main arboviruses in the world, exposing more than half of the population to the risk of contracting the disease [1–3]. In addition, the severe dengue is one of the main causes of child deaths in countries in Latin America and Asia [3]. The circulation of the dengue virus is ruled as a major problem due to the genetic plasticity of the mosquito under environmental conditions, and due to the selection of resistant insects because of chemical control, coupled with the fact that the mosquito is anthropophilic and domiciled. The occurrence of vertical transmission in epidemic times also helps in the dynamics of maintaining the virus in the urban environment [4]. Murillo et al. [5] reported that increased vertical transmission in times of epidemic makes outbreaks more difficult and expensive to control.

Aedes (*Stegomyia*) *aegypti* (Diptera: Culicidae) is the transmitter of dengue, urban yellow fever, chikungunya and Zika viruses in Brazil [6, 7]. The spread of chikungunya and Zika viruses has caused public health problems because the former causes severe recurrent joint pain and the latter may involve neurological complications such as microcephaly in children, being very severe, in addition to Guillain-Barré Syndrome [8–11].

Females of this mosquito species practice hematophagy and prefer to breed in artificial breeding sites, where competition with other species is practically absent [12–15]. The proliferation of this mosquito has been intensifying with increasing urban perimeter, lack of basic sanitation, deforestation and lack of proper treatment of solid waste, as these are factors that increase breeding sites [16, 17]. *A. aegypti* eggs can be dehydrated for more than a year and are viable in late-hatching post-embryonic development, causing an imminent danger in the proliferation of this mosquito [18, 19].

Mosquito development, especially those with high adaptive genetic plasticity, may be affected by various abiotic factors, with temperature being the most relevant [20]. According to the Intergovernmental Panel on Climate Change [21], the Earth will experience an average temperature increase of approximately 2°C, and by 2099 this could rise to 6.5°C, a situation that may favor the cycle of these Culicidae, as well as increased viral circulation. These data are confirmed by analyzing the statistics from 1955 to 2007, a time when temperature increases have been added to one of the world's largest viral propagations, with a 30-fold increase in the last 50 years, and over 2 million of reported cases annually on the American continent each year [3].

In the current perspective, there is a need to intensify the monitoring and control of mosquitoes, as well as the monitoring of early viral circulation, since almost half of the world's population is exposed to the risk of dengue contamination with an estimated 50–100 million cases annually in more than 100 endemic countries [3]. This situation can worsen with global warming, which has been increasing each year, due to anthropic factors [17]. In this sense, understanding the life cycle of *A. aegypti* in stabilized insectaries kept in environmental rooms simulating the IPCC climate predictions, is a strategy that can help to optimize the alternatives for monitoring and controlling of this insect, given the possibility of the emergence of this scenario.

In this context, the present work aimed to: (i) verify the influence of different temperatures on egg viability, number of adults and mortality rate of *A. aegypti* from Londrina, Paraná, under laboratory conditions and, (ii) check the time of the biological cycle, number of adults and mortality rate of *A. aegypti* from Manaus, Amazonas, kept in environmental rooms that simulate the temperatures and concentrations of carbon dioxide predicted by the IPCC.

2. Methods

2.1 Study area

The study was carried out in incubator chambers (BOD) with different temperatures, kept at the Medical Entomology Laboratory of Londrina State University (UEL), Londrina, Paraná, and in insects kept at the Laboratory of Ecophysiology and Molecular Evolution (LEEM), located at *Campus* I of the National Institute of Amazonian Research (INPA), Manaus, Amazonas.

2.2 Collection of *A. aegypti* eggs from Londrina, Paraná, under field conditions

The sample of *A. aegypti* population from Londrina, Paraná, used in this study, was obtained from eggs collected in the field, with the aid of traps called ovitraps [22]. The main attraction in the traps was a solution containing grass infusion (10%) [23], with a total volume of 300 ml. Twenty traps were set up and distributed at the UEL campus, at ground level, protected from sun and rain, and in places with little movement of people and animals. Samples were taken for 2 weeks and every 7 days, the reeds were replaced and sent to the insectarium with controlled temperature, humidity and photoperiod conditions (27 ± 2°C, 80–90% and 12L/12E).

2.3 Maintenance of immature and adult of *A. aegypti* from Londrina, Paraná, under laboratory conditions

The reeds containing eggs were immersed in plastic trays (45 × 30 × 7.5 cm) containing distilled water to stimulate the larvae to hatch. The immatures obtained were kept until adults by means of food containing a mixture of cat food (Whiskas®) and rodents (Teklab global®) in a 1:1 ratio, ground into fine particles (1 mm). All trays were covered with a nylon fabric to prevent the escape of mosquitoes. After emergence, the males and females were collected with a Castro catcher and the species identification was performed using external morphological characters, mainly from the chest, with the aid of a stereoscopic microscope ZEISS Stemi 2000 50× and identification keys proposed by Forattini [14], Harbach [24], and WRBU [25].

After the identification stage, the adults were placed to copulate in cardboard cages (17 × 20 cm), containing two plastic cups lined with strips of filter paper and filled with 70 mL of distilled water, which were used as a substrate for oviposition. As a source of carbohydrates, an Erlenmeyer was introduced containing a roll of gauze with pieces of cotton in the center, soaked with 12% sugar water. The blood meal was carried out using an anesthetized hamster for 30min, according to the procedure approved by the Ethics Committee on the Use of Animals at UEL ("Breeding of mosquitoes in laboratory conditions").

2.4 Biological aspects of *A. aegypti* from Londrina, Paraná, in incubator chambers (BOD) with different temperatures

At UEL, the experiment was carried out in four incubator chambers (BOD) that had different temperatures (0, 5, 25 and 45°C) with ±2°C, as well as at ambient temperature, where the eggs remained in dry incubation in 500 ml capacity pots containing only moistened cotton over a period of 10 days, with a 12/12 h photoperiod.

Subsequently, 150 ml of distilled water, 25 eggs of *A. aegypti*, F1 field generation and 0.055 g of larval food were introduced into each of the five pots, which were placed in BOD with temperature of 25 ± 2°C. The monitoring of the hatching rate of eggs, the rate of immature deaths and the number of adults were carried out daily.

2.5 Stabilization of a sample of the population of *A. aegypti* from Manaus, Amazonas, in environmental rooms with temperature and CO_2 conditions according to the provisions of the IPCC

The same process of creating and maintaining the immature and adults of *A. aegypti* from Londrina, Paraná, mentioned in items 2.2 and 2.3, was carried out with a sample of the population of *A. aegypti* from Manaus, Amazonas, collected in the same way on *Campus* I of INPA. However, the blood meal carried out followed the procedure approved by the Ethics Committee on the Use of Animals at INPA (CEUA/INPA: 04/2013-"Breeding of mosquito vectors, under laboratory conditions").

After the stabilization of the three insectaries in the LEEM, the F28 generation was obtained in the different environmental rooms that started the experiment for post-embryonic development, using three different environments. Rooms 1, 2 and 4 suffer from interference from different temperatures and levels of carbon dioxide, and rooms 2 and 4 refer to 50 and 100 years in front of room 1, which presents ambient temperature.

These rooms are equipped with technological devices that guarantee, respectively, the maintenance of ambient temperature, 2.5°C (400 ppm CO_2) and 4.5°C (850 ppm CO_2) above ambient temperature. The natural conditions of the respective environmental rooms are collected in real time by sensors isolated in the forest and the humidity of all remains constant in relation to the environmental conditions.

2.6 Biological aspects of *A. aegypti* from Manaus, Amazonas, in insectaries kept in environmental rooms that present simulated conditions of climate forecast by the IPCC

After stabilization of insectaries in environmental rooms 1 (ambient temperature), 2 (mild temperature) and 4 (extreme temperature), 125 eggs from each room were counted with the aid of a stereoscopic microscope. These eggs were separated into 5 groups (replicates) of 25 eggs, kept in pots with capacity of 750 mL, containing 300 mL of distilled water and 0.055 g of larval food added every 3 days.

The hatching rate was monitored daily for 19 days under these conditions. The average time of the biological cycle, number of adults (males and females) and larval mortality rate were verified. Exuvia and dead immatures were removed from the pots and the volume of distilled water was replaced.

2.7 Statistical analysis

Data on the hatching rate of eggs, number of adults (males and females) and mortality rate of immature *A. aegypti* from Londrina, Paraná, were expressed as a percentage. The number of emerged adults was subjected to a Kruskal-Wallis test, at the level of 5% significance, in order to verify differences between the data in the temperatures analyzed, using the BioEstat® 5.3 program for Windows [26].

The results on the number of adults (males and females) and the larval mortality rate of *A. aegypti* from Manaus, Amazonas, were also analyzed by means of percentage. On the other hand, the data regarding the average time of the biological cycle of mosquitoes kept in the different environmental rooms, were analyzed using averages, standard deviation and standard error. At first, these data underwent a Lilliefors normality test (K samples) to find out whether they have a normal distribution or not, and subsequently subjected to Analysis of Variance (ANOVA), followed by Tukey's multiple comparison test ($p \leq 0.05$), with the aid of the statistical program SPSS® 14.0 packpage for Windows® (SPSS Inc. 2005 Headquarters, Chicago, Illinois, USA).

3. Results

3.1 Viability of eggs and total adults of *A. aegypti* from Londrina, Paraná, in incubator chambers (BOD) with different temperatures

The highest hatching rates of eggs were obtained at temperatures of 5 and 25 ± 2°C, respectively, where 48% for both was verified, followed by ambient temperature with 36% and at 0 ± 2°C presenting 7%. On the other hand, there was no hatching of eggs at 45 ± 2°C (**Figure 1**).

The adults obtained in the proportion of males and females are shown in **Table 1**.

At 0 ± 2°C, seven of nine immatures reached adulthood, which indicates a mortality of 22.2% of the larvae at this temperature. Considering the 60 immatures obtained at a temperature of 5 ± 2°C, 56 reached the adult stage indicating a larval mortality of 6.6%, while at a temperature of 25 ± 2°C only 50 immatures among 60 reached the adult stage, indicating a larval mortality of 16.6%. At ambient temperature, it was observed that of 45 immatures obtained, 44 reached adulthood, representing a larval mortality of only 2.2%. On the other hand, there was no adult emergency at 45 ± 2°C.

There is no statistical difference between hatching rates at 0, 5, 25 and ambient temperatures (**Table 1**), while at 0°C there is difference in relation to the hatching rate at other temperatures.

Of the total 625 eggs incubated at different temperatures, 157 eggs reached adulthood, and of these, 98 were males and 59 females, making up a male:female ratio of 1.66:1. The proportion of emerged adults was 5.6% at 0°C, 44.8% at 5°C, 40% at 25°C and 35.2% at room temperature (**Table 1**).

3.2 Average biological cycle time and total adults of *A. aegypti* from Manaus, Amazonas, in environmental rooms that simulate the climatic conditions provided by the IPCC

The averages of climatic variations during the experiment in °C, increase in ppm of CO_2 and percentage (%) of humidity for the different environmental rooms are represented in **Table 2**.

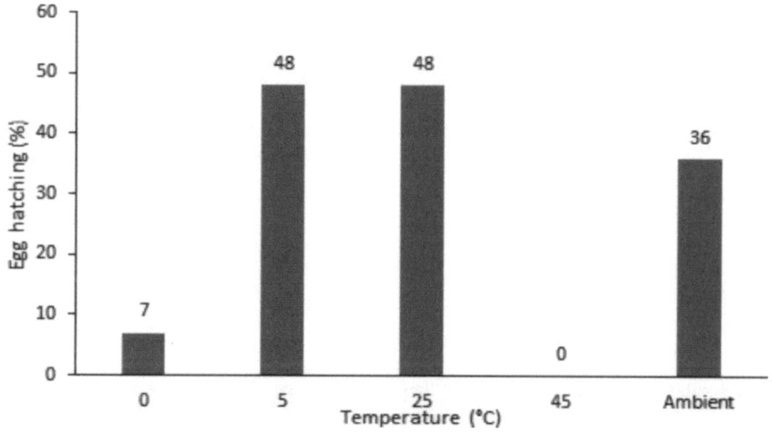

Figure 1.
Egg hatch percentage of A. aegypti *from Londrina, Paraná, in incubator chambers (BOD) with different temperatures with ±2°C limits and ambient temperature (16.7–24.1°C), observed for 10 days.*

Adult emergence rate	Temperature (±2°C)									
	0°C		5°C		25°C		45°C		Ambient (16.7–24.1°C)	
Replicas	M	F	M	F	M	F	M	F	M	F
1	4	1	8	7	10	4	0	0	6	5
2	0	1	5	4	8	6	0	0	0	0
3	0	0	8	4	5	2	0	0	0	0
4	0	1	9	7	12	3	0	0	10	6
5	0	0	4	0	0	0	0	0	9	8
Total	4	3	34	22	35	15	0	0	25	19
Total M-F*	7a⁺		56a		50a		0b		44a	
Total %	5.60%		44.80%		40%		0%		35.20%	

*M-F: M = male and F = female.

⁺*Equal letters on the same line do not differ from each other by the Kruskal-Wallis $p = 0.0261$, $H = 11.0428$.*

Table 1.
Emergence of Aedes aegypti *adults with eggs incubated for 10 days at different temperatures and placed for post-embryonic development at a temperature of 25 ± 2°C.*

Rooms	Climate variations		
	Temperature (°C)	CO_2 (ppm)	Humidity (%)
1	27.23	434.4	77.01
2	29.74	864.6	78.66
4	32.18	1279.87	76.96

Table 2.
Averages of temperature °C, increase in ppm of CO_2 and percentage (%) of humidity for the different environmental rooms.

There was a statistical difference ($p < 0.05$) between the average time of the biological cycle of *A. aegypti* in the different environments, with room 4 being the environment that had the greatest influence on the life cycle of mosquitoes, since they developed more faster than the insects kept in the other environmental rooms (**Table 3**).

In the room 1, it was found that adults emerged on the sixth and seventh day, with 45.79% females and 54.20% males. In the room 2, adults emerged on the sixth day, with 58.88% of females and 41.11% of males. In the room 4, adults began to emerge on the fifth and sixth day of the experiment, with 60.34% females and 39.66% males. The mortality percentage in the different instars for rooms 1, 2 and 4 was: 14.4; 28 and 53.6%, respectively.

4. Discussion

Temperature is one of the main environmental variables responsible for changes in the biology of the mosquito *A. aegypti*, directly influencing the reproductive cycle of this species [27–31]. The exposure of *A. aegypti* eggs to temperatures out of their normal range can cause physiological stress, interrupting the development [32]. Therefore, the temperature must be in a suitable range, in order to the larvae hatch and the generations develop and multiply.

Parâmeters	Room 1 (TA)	Room 2 (TM)	Room 4 (TE)
Averages	7.3430[a,*]	6.6780[b,*]	6.0510[c,*]
Standard deviation	0.43756	0.65009	0.60656
Standard error	0.13837	0.20558	0.19181

TA = ambient temperature; TS = mild temperature; TE = extreme temperature. For the Tukey test, the averages were transformed into a root of x + 0.5. Different letters on the same line, indicate a statistically significant difference according to the Tukey test (p < 0.05), CV = 4%; F = 12.64 and DMS = 0.120.*

Table 3.
Average life cycle time of adults kept in different environmental rooms that simulate the climatic conditions provided by the IPCC.

In this study, the high hatching rate of larvae (48%) maintained at a temperature of 25°C, in Londrina, Paraná, corroborates what was observed in the laboratory by Farnesi et al. [33] and Mohammed and Chadee [34], who verified larvae hatching rates above 90% at this temperature. In addition, it is in accordance with the optimal temperature range for the development of the *A. aegypti* mosquito, mentioned in the literature.

Beserra et al. [35] observed that this range is between 21 and 29°C, when studying the thermal requirements of the species in four bioclimatic regions of Paraíba, Brazil. Later, Beserra et al. [36] observed that the optimal temperature for the development of *A. aegypti* is between 22 and 32°C, when studying the mosquito's biological cycle in air-conditioned chambers with six different temperatures. For Marinho [37], the ideal temperature range for the development of the vector is between 22°C and 36°C, according to results obtained after analyzing the influence of climatic factors on the pattern of oviposition, distribution and population development of *A. aegypti* in the field and the laboratory. More recently, Galavíz-Parada et al. [29] conclude that hatching and survival of *A. aegypti* in Mexico can occur in a temperature range between 15°C and 32°C under laboratory conditions.

Thus, the third highest rate of hatching of larvae (35%), as well as the lowest mortality of larvae (2.2%) occurring at ambient temperature (16.7–24.1°C) in Londrina, is also in line with the range optimal temperature for the development of *A. aegypti*, mentioned in the literature, and corroborates with a study by Yang et al. [38], where they observed that the mortality rate of adult females of *A. aegypti* was lower between 15°C and 30°C and increases rapidly at temperatures below or above this range, thus corroborating with the mortality rate observed at 0°C (22.2%) in Londrina. On the other hand, Costa et al. [39] reported that the longevity in *A. aegypti* decreased from 25 to 35°C, corroborating with the second highest mortality rate, observed at 25° C (16.6%) in Londrina.

Assessing the influence of ambient temperature on the longevity and fertility of *A. aegypti* in the city of Guarapuava, Paraná, Ajuz et al. [40] found that the vector's survival zone is wide and that only the minimum temperatures below 5°C limit the proliferation and super infestation of the species in the city. Thus, it is evident that the life cycle of the mosquitoes in Londrina is also affected only in extreme temperatures, in view of the greater number of emergencies of adults of *A. aegypti* observed at 5°C (44.80%), as well as a relatively low mortality rate on this temperature (6.6%). The ability of *A. aegypti* larvae and adult to tolerate low temperatures also was demonstrated in the study by Jass et al. [41].

The low hatching of larvae (7%) and emergence of adults (5.6%) at 0°C, as well as the absence at 45°C observed in Londrina, is justified by the fact of extreme temperatures (very low or very high) are harmful to the development of *A. aegypti*, according to Buriol et al. [42], who stated that temperatures below 5°C and above 40°C are lethal to mosquito development. Similar data were observed by De Majo et al. [43], Marinho et al. [44], Mohammed and Chadee [34] in studies about the effect of different temperatures on the development of *A. aegypti*. More recently, Sukiato et al. [32] also noted in their study that there was no hatching of *A. aegypti* eggs at 40°C, and the larval and pupal mortality was higher at 37°C, compared to other lower temperatures (34, 31 and 28°C).

In Manaus, Amazonas, the greatest influence of the room 4 (with the highest temperature - 32.18°C) on the life cycle of the mosquitoes, where they developed faster than mosquitoes kept in other lower temperature environmental rooms, is in line with Carrington et al. [45], who found that temperatures around 30°C are ideal for the development of *A. aegypti* and that the development was faster under a temperature of 35°C and impaired above this range. Similar results were also obtained by Brady et al. [46], when evaluating the survival of the species at different temperatures, located between 0°C and 40°C. Sukiato et al. [32] also observed that the *A. aegypti* development time was shorter at higher temperatures (37 and 34°C).

In relation to the mortality rate, the lower mortality (14.4%) that occurred at the room 1, where the temperature was lower, as well as the higher mortality (53.6%) of different instars that occurred at the room 4, also corroborate with the results reported by Yang et al. [38] and Sukiato et al. [32] already compared with the results obtained in Londrina. A study by Farjana et al. [47] also demonstrated that the mortality of *A. aegypti* increased at a higher temperature (35°C).

The lowest and the highest mortality rate at the rooms 1 and 4, respectively, may have happened due to the influence of CO_2 concentration—which was lower at the room 1 (similar to the current atmospheric concentration) and much higher at the room 4—because CO_2 atmospheric is also related to the biological cycle of living beings, influencing their ecological interactions. High CO_2 rates have an impact on ecological communities, causing a reduction in nitrogen concentration. Furthermore, it can reduce the quality and quantity of food in breeding sites, compromising larval growth and survival [48].

However, in a study by Azevedo et al. [49], higher concentrations of CO_2 had no significant influence on the results obtained in relation to the biological cycle of *A. aegypti*.

5. Conclusion

In the South region of the country, the hatching of eggs and emergence of adults of *A. aegypti* only did not occur at 45°C, while the lowest rate of hatching and emergence occurred at 0°C, indicating that the development of the mosquitoes in Londrina is affected only in very extreme temperatures, since the temperature of 5°C still proved beneficial to the development of *A. aegypti*. In the North region of the country, the development of the immature in Manaus is faster as the temperature in the environmental rooms increases; however, at the same time, death rates also increase.

Therefore, it is concluded that temperatures from 5 to 29.74°C are more appropriate, since values outside these limits can cause deleterious effects in biological aspects related to the reproductive success of the species. Thus, temperature has a great influence on these aspects, with medium temperatures being more beneficial to this species.

The results obtained show that both *A. aegypti* mosquitoes from the South and North regions of Brazil have adaptive potential in face of the increase in the average temperature stipulated by the IPCC, in view of the unviability of eggs only at extreme temperatures and considering the shorter average duration of the life cycle observed at high temperatures. Thus, the predicted climate changes may favor the development and proliferation of *A. aegypti*, and consequently the viral circulation, in addition to make possible the occurrence of a geographical expansion of *A. aegypti*. However, it is important to mention that other environmental variables can also influence the biology of mosquitoes, as well as viruses, requiring, therefore, more studies related to the various environmental variables and viruses in order to be able to affirm whether there will be a greater occurrence of arboviruses due to global warming.

Acknowledgements

We are grateful to the technicians of the Malaria and Dengue Laboratories (INPA) of the Medical Entomology Laboratory (UEL). This work is supported by the State University of Londrina and ADAPTA/CNPq: 573976/2008-2.

Conflict of interest

The authors declare no conflict of interest.

Author details

Ediane Oliveira do Amaral[1], André Felipe da Silva[2], William Ribeiro da Silva[1,3], Karina Rossi da Silva[2], Rosemary Aparecida Roque[1], Wanderli Pedro Tadei[1,3] and João Antonio Cyrino Zequi[2]*

1 Laboratório de Controle Biológico e Biotecnologia da Malária e Dengue, Instituto Nacional de Pesquisas da Amazônia, Avenida André Araújo, Petrópolis, Manaus, Amazona, Brazil

2 Laboratório de Entomologia Médica, Departamento de Biologia Animal e Vegetal, Universidade Estadual de Londrina, Programa de Pós-Graduação em Ciências Biológicas, Londrina, Paraná, Brazil

3 Programa de Pós-Graduação em Entomologia, Instituto Nacional de Pesquisas da Amazônia, Manaus, Amazonas, Brazil

*Address all correspondence to: zequi@uel.br

References

[1] Sarwar M. Proposals for the control of principal dengue fever virus transmitter *Aedes aegypti* (Linnaeus) mosquito (Diptera: Culicidae). Journal of Ecology and Environmental Sciences. 2014;**2**(2):24-28

[2] Sarwar M. Reducing dengue fever through biological control of disease carrier *Aedes* mosquitoes (Diptera: Culicidae). International Journal of Preventive Medicine Research. 2015;**1**(3):161-166

[3] World Health Organization (WHO). 2020. Available from: https://www.who.int/en/news-room/fact-sheets/detail/dengue-and-severe-dengue [Accessed: 27 May 2020]

[4] Martins VEP, Alencar CH, Kamimura MT, de Carvalho Araujo FM, De Simone SG, Dutra RF, et al. Occurrence of natural vertical transmission of dengue-2 and dengue-3 viruses in *Aedes aegypti* and *Aedes albopictus* in Fortaleza, Ceará, Brazil. PLoS One. 2012;**7**(7):e41386

[5] Murillo D, Murillo A, Lee S. The role of vertical transmission in the control of dengue fever. International Journal of Environmental Research and Public Health. 2019;**16**(5):803

[6] Leta S, Beyene TJ, De Clercq EM, Amenu K, Kraemer MU, Revie CW. Global risk mapping for major diseases transmitted by *Aedes aegypti* and *Aedes albopictus*. International Journal of Infectious Diseases. 2018;**67**:25-35

[7] Vega-Rúa A, Zouache K, Girod R, Failloux AB, Lourenço-de-Oliveira R. High level of vector competence of *Aedes aegypti* and *Aedes albopictus* from ten American countries as a crucial factor in the spread of Chikungunya virus. Journal of Virology. 2014;**88**(11):6294-6306

[8] Malta JMAS, Vargas A, Leite PL, Percio J, Coelho GE, Ferraro AHA, et al. Síndrome de Guillain-Barré e outras manifestações neurológicas possivelmente relacionadas à infecção pelo vírus Zika em municípios da Bahia, 2015. Epidemiologia e Serviços de Saúde. 2017;**26**:9-18

[9] Nóbrega MEBD, Araújo ELDL, Wada MY, Leite PL, Dimech GS, Pércio J. Surto de síndrome de Guillain-Barré possivelmente relacionado à infecção prévia pelo vírus Zika, Região Metropolitana do Recife, Pernambuco, Brasil, 2015. Epidemiologia e Serviços de Saúde. 2018;**27**:e2017039

[10] Secretaria de Vigilância em Saúde (SVS)/Ministério da Saúde (MS). Monitoramento dos casos de arboviroses urbanas transmitidas pelo *Aedes* (Dengue, Chikungunya e Zika) até a Semana Epidemiológica 11 de 2019. Boletim Epidemiológico. 2019;**50**. Available from: http://portalarquivos2.saude.gov.br/images/pdf/2019/marco/25/2019-013Monitoramento-dos-casos-de-arboviroses-publicacao-25-03-2019.pdf [Accessed: 22 April 2019]

[11] World Health Organization (WHO). 2016. Available from: https://www.who.int/csr/don/8-february-2016-gbs-brazil/en/ [Accessed: 15 February 2019]

[12] Carvalho FD, Moreira LA. Why is *Aedes aegypt*i Linnaeus so successful as a species? Neotropical Entomology. 2017;**46**(3):243-255

[13] Consoli RAGB, Oliveira RL d. Principais mosquitos de importância sanitária no brasil. Rio de Janeiro: Fio Cruz; 1994. pp. 228

[14] Forattini OP. Culicidologia Médica. São Paulo: Editora USP, São Paulo-SP; 2002. pp. 860

[15] Lourenço-de-Oliveira R. Biologia e comportamento do vetor. In: Valle D, Pimenta DN, Cunha RV, editors. Dengue teorias e práticas. Rio de Janeiro: Fiocruz; 2015. pp. 76-92

[16] Campanelli ES. O desenvolvimento de um processo de infecção do *Aedes aegypti* pelo vírus dengue: caracterização da interação do vírus com uma população de mosquitos autóctones. Dissertação (Mestrado em Ciências da Saúde). Belo Horizonte: Fundação Oswaldo Cruz "Centro de Pesquisas René Rachou"; 2007. pp. 91

[17] Dickens BL, Sun H, Jit M, Cook AR, Carrasco LR. Determining environmental and anthropogenic factors which explain the global distribution of *Aedes aegypti* and *Ae. albopictus*. BMJ Global Health. 2018;**3**(4):e000801

[18] Liu-Helmersson J, Stenlund H, Wilder-Smith A, Rocklöv J. Vectorial capacity of *Aedes aegypti*: Effects of temperature and implications for global dengue epidemic potential. PLoS One. 2014;**9**(3):e89783

[19] Soares-Pinheiro VC, Dasso-Pinheiro W, Trindade-Bezerra JM, Tadei WP. Eggs viability of *Aedes aegypti* Linnaeus (Diptera, Culicidae) under different environmental and storage conditions in Manaus, Amazonas, Brazil. Brazilian Journal of Biology. 2017;**77**(2):396-401

[20] Kamal M, Kenawy MA, Rady MH, Khaled AS, Samy AM. Mapping the global potential distributions of two arboviral vectors *Aedes aegypti* and *Ae. albopictus* under changing climate. PLoS One. 2018;**13**:e0210122

[21] IPCC. IPCC Expert Meeting: Climate research community looks into future scenarios. 2007. Available from: www.ipcc.ch/news_and_events/pdf/press/180512_pr_Expert_Meeting_scenarios.pdf [Accessed: 30 March 2020]

[22] Fay RW, Eliason DA. A preferred oviposition site as a surveillance method for *Aedes aegypti*. Mosquito News. 1966;**26**:531-534

[23] Reiter P, Amador MA, Colon N. Enhancement of the CDC ovitrap with hay infusions for daily monitoring of *Aedes aegypti* populations. Journal of the American Mosquito Control Association. 1991;7:52-55

[24] Harbach RE. Mosquito Taxonomic Inventory. 2020. Available from: http://mosquito-taxonomic-inventory.info/ [Accessed: 30 March 2020]

[25] Walter Reed Biosystematics Unit (WRBU). Mosquito identification resources. 2020. Available from: http://www.wrbu.org/VecID_MQ.html [Accessed: 21 January 2020]

[26] Ayres M, Ayres J, Ayres DL, Santos AS. BioEstat versão 5.3: Aplicações estatísticas nas áreas das ciências Biológicas e médicas. Belém/Belém, Pará, Brasil: Sociedade Civil Mamirauá, Brasília/MCT/CNPQ; 2007

[27] Couret J, Benedict MQ. A meta-analysis of the factors influencing development rate variation in *Aedes aegypti* (Diptera: Culicidae). BMC Ecology. 2014;**14**(1):3

[28] Galati EA, Tamara NDL, Natal D, Chiaravalloti-Neto F. Mudanças climáticas e saúde urbana. Revista USP. 2015;**107**:79-90

[29] Galavíz-Parada J, Vega-Villasante F, Marquetti MDC, Guerrero-Galván S, et al. Efecto de la temperatura y salinidad en la eclosión y supervivencia de *Aedes aegypti* (L)(Diptera: Culicidae) procedentes del occidente de México. Revista Cubana de Medicina Tropical. 2019;**71**(2):e353

[30] Goindin D, Delannay C, Ramdini C, Gustave J, Fouque F. Parity and longevity of *Aedes aegypti* according

to temperatures in controlled conditions and consequences on dengue transmission risks. PLoS One. 2015;**10**(8):e0135489

[31] Mbaika S, Lutomiah J, Chepkorir E, Mulwa F, Khayeka-Wandabwa C, Tigoi C, et al. Vector competence of *Aedes aegypti* in transmitting Chikungunya virus: Effects and implications of extrinsic incubation temperature on dissemination and infection rates. Virology Journal. 2016;**13**(1):114

[32] Sukiato F, Wasserman RJ, Foo SC, Wilson RF, Cuthbert RN. The effects of temperature and shading on mortality and development rates of *Aedes aegypti* (Diptera: Culicidae). Journal of Vector Ecology. 2019;**44**(2):264-270

[33] Farnesi LC, Martins AJ, Valle D, Rezende GL. Embryonic development of *Aedes aegypti* (Diptera: Culicidae): Influence of different constant temperatures. Memórias do Instituto Oswaldo Cruz. 2009;**104**(1):124-126

[34] Mohammed A, Chadee DD. Effects of different temperature regimens on the development of *Aedes aegypti* (L.) (Diptera: Culicidae) mosquitoes. Acta Tropica. 2011;**119**(1):38-43

[35] Beserra EB, Castro FP Jr, Santos JW, Santos TS. Biologia e exigências térmicas de *Aedes aegypti* (l.) (Diptera: Culicidae) provenientes de quatro regiões bioclimáticas da Paraíba. Neotropical Entomology. 2006;**35**:853-860

[36] Beserra EB, Fernandes CR, Silva SADO, Silva LAD, Santos JWD. Efeitos da temperatura no ciclo de vida, exigências térmicas e estimativas do número de gerações anuais de *Aedes aegypti* (Diptera, Culicidae). Iheringia. Série Zoologia. 2009;**99**(2):142-148

[37] Marinho RA. Ecobiologia de *Aedes aegypti* (L. 1762) (Diptera: Culicidae) associada a fatores climáticos em três mesorregiões da Paraíba. 2013

[38] Yang HM, Macoris MDLDG, Galvani KC, Andrighetti MTM, Wanderley DMV. Assessing the effects of temperature on the population of *Aedes aegypti*, the vector of dengue. Epidemiology & Infection. 2009;**137**(8):1188-1202

[39] Costa EAPDA, Santos EMDM, Correia JC, Albuquerque CMRD. Impact of small variations in temperature and humidity on the reproductive activity and survival of *Aedes aegypti* (Diptera, Culicidae). Revista Brasileira de Entomologia. 2010;**54**(3):488-493

[40] Ajuz LC, Vestena LR. Influência da pluviosidade e temperatura ambiente na longevidade e fecundidade dos *Aedes aegypti* e *albopictus* na cidade de Guarapuava-PR e possibilidade de superinfestação. Hygeia. 2014;**10**(18):1-18

[41] Jass A, Yerushalmi GY, Davis HE, Donini A, MacMillan HA. An impressive capacity for cold tolerance plasticity protects against ionoregulatory collapse in the disease vector *Aedes aegypti*. Journal of Experimental Biology. 2019;**222**(24):jeb214056

[42] Buriol GA, Estefanel V, Gracioli MDSA, Fantineli DG, de Chagas ÁC. Zoneamento climático das condições para o desenvolvimento da larva do mosquito transmissor do vírus da dengue no Estado do Rio Grande do Sul. Revista Eletrônica de Comunicação, Informação e Inovação em Saúde. 2009;**3**(2):24-36

[43] De Majo MS, Montini P, Fischer S. Egg hatching and survival of immature stages of *Aedes aegypti* (Diptera: Culicidae) under natural temperature conditions during the cold season in Buenos Aires, Argentina. Journal of Medical Entomology. 2017;**54**(1):106-113

[44] Marinho RA, Beserra EB, Bezerra-Gusmão MA, Porto VDS,

Olinda RA, dos Santos CA. Effects of temperature on the life cycle, expansion, and dispersion of *Aedes aegypti* (Diptera: Culicidae) in three cities in Paraiba, Brazil. Journal of Vector Ecology. 2016;**41**(1):1-10

[45] Carrington LB, Armijos MV, Lambrechts L, Barker CM, Scott TW. Effects of fluctuating daily temperatures at critical thermal extremes on *Aedes aegypti* life-history traits. PLoS One. 2013;**8**(3):e58824

[46] Brady OJ, Johansson MA, Guerra CA, Bhatt S, Golding N, Pigott DM, et al. Modelling adult *Aedes aegypti* and *Aedes albopictus* survival at different temperatures in laboratory and field settings. Parasites & Vectors. 2013;**6**(1):351

[47] Farjana T, Tuno N, Higa Y. Effects of temperature and diet on development and interspecies competition in *Aedes aegypti* and *Aedes albopictus*. Medical and Veterinary Entomology. 2012;**26**(2):210-217

[48] Smith C, Baldwin AH, Sullivan J, Leisnham PT. Effects of elevated atmospheric CO_2 on competition between the mosquitoes *Aedes albopictus* and *Ae. triseriatus* via changes in litter quality and production. Journal of Medical Entomology. 2013;**50**(3):521-532

[49] Azevedo JB. Análise do ciclo biológico do *Aedes aegypti* (Diptera: Culicidae) exposto a cenários de mudanças climáticas previstas pelo IPCC (Intergovernmental Panel on Climate Change) [Dissertação (Mestrado)]. Manaus: INPA; 2015. pp. 53

Chapter 7

Bioecology of Blossom Midge of Jasmine, *Contarinia maculipennis* Felt (Cecidomyiidae: Diptera) in Different *Jasminum* Cultivars

I. Merlin Kamala

Abstract

Jasmine is a well-renowned flower around the world especially in the tropics because of its unique fragrance. *Jasminum sambac, Jasminum auriculatum, Jasminum grandiflorum,* and *Jasminum nitidum* are the four cultivable species of *Jasminum.* Blossom midge, *Contarinia maculipennis* Felt (Cecidomyiidae, Diptera), a dipteran pest, has attained the status of a major pest causing severe economic loss by reducing the marketable quality of the flowers. To compare the lifecycle of the notorious midge in all the cultivable *Jasminum* species, a detailed laboratory study was conducted at the Department of Agricultural Entomology, Tamil Nadu Agricultural University, Coimbatore, India. The observations on life cycle parameters of blossom midge, *C. maculipennis*, revealed that the life cycle consists of egg, maggot, pupa, and adult. The life cycle and duration of each stage are found to be the shortest on *J. nitidum* and the longest on *J. sambac.*

Keywords: *Jasminum sambac*, *Jasminum auriculatum*, *Jasminum grandiflorum*, *Jasminum nitidum*, blossom midge

1. Introduction

The genus *Jasminum*, belonging to Oleaceae or Olive family, contains over 200 species of vines or shrubs with opposite leaves and fragrant flowers [1]. The name "jasmine" or "jessamine" is derived from the Arabic and Persian "yasmin" or "yasaman" and originally referred to the species of Arabian jasmine (*Jasminum sambac* (L.) Ait) [2]. This plant, with its extremely fragrant flowers, is native to southern Asia, but it has been cultivated in the Middle East for centuries [3]. *Jasminum* is the generic name of shrubs and vines in the olive family (Oleaceae). Although more than 200 species are known, 40 species have been identified in India, and 20 species are cultivated in South India [4–6], of which only three species are used for commercial cultivation namely, *J. sambac* (Gundumalli/Madurai Malli), *Jasminum auriculatum* (Mullai), and *Jasminum grandiflorum* (Jathimalli/Pitchi). The angel jasmine, *Jasminum nitidum*, with sweetly fragrant, snow-white, pinwheel-shaped flowers is recently introduced for commercial cultivation in Tamil Nadu, India.

2. *Jasminum sambac*

J. sambac is an evergreen vine or shrub reaching up to 0.5–3 m (1.6–9.8 ft) tall. The plant produces flowers all throughout the year in clusters of 3–12 together at the ends of branches. They are strongly scented flowers with a white corolla. The flowers open at night (usually around 6–8 in the evening). The sweet, heady fragrance of *J. sambac* is its distinct feature (**Figure 1**). It is widely grown throughout the tropics from the Arabian Peninsula to Southeast Asia and the Pacific Islands as an ornamental plant and for its strongly scented flowers [7].

***J. sambac* plant** | ***J. sambac* flower** | ***J. sambac* flower bud**

Figure 1.
Jasminum sambac.

3. *Jasminum auriculatum*

J. auriculatum is a species of jasmine, in the family Oleaceae. It is found in India and other Asian countries. It is cultivated commercially in India and Thailand for its commercial oil. It is used for decorative purposes in occasions and festivals in India. It is a small climbing bushy plant with simple ovate dark green small leaves and small white flowers. Leaves are opposite, sometimes hairless, simple, or trifoliolate. Lateral leaflets are much smaller, rarely exceeding 4 mm in diameter, the central one up to 3.5 cm long and 1.5 cm broad, ovate, shortly pointed. Flowers are fragrant in many-flowered cymes. Flower stalks are up to 5 mm long. Flowers are white, tube 1.5 cm long, lobes elliptic, up to 8 mm long. Berry is 5 mm in diameter, globose, black (**Figure 2**) [8].

***J. auriculatum* plant** | ***J. auriculatum* flower** | **Flower bud**

Figure 2.
Jasminum auriculatum.

4. *Jasminum grandiflorum*

J. grandiflorum is also known variously as the Spanish jasmine, Royal jasmine, and Catalan jasmine. It is a stunning deciduous shrub growing to 2–4 m tall.

***J. grandiflorum* plant** **_J. grandiflorum_ flower** **Flower bud**

Figure 3.
Jasminum grandiflorum.

The leaves are opposite, 5–12 cm long, with 5–11 leaflets in pinnate form. The flowers are produced in open cymes, and the individual flowers are white having corolla with a basal tube 13–25 mm long and five lobes 13–22 mm long. The flower's fragrance is unique and sweet. It is widely cultivated as an ornamental plant in warm temperate and subtropical regions. The Jasmine Concrete and Jasmine Oleoresin (sold as Jasmine Absolute) were extracted by solvent extraction method. Both products have a huge demand in the fragrance industry (**Figure 3**). Methyl jasmonate isolated from the jasmine oil of *J. grandiflorum* has led to the discovery of the molecular structure of the jasmonate plant hormones [9].

5. *Jasminum nitidum*

This twinning climber has slender stems and forms a dense habit. The dark glossy green leaves are lance shaped, and the fragrant white star-shaped flowers appear throughout summer in clusters (**Figure 4**). The angle-winged jasmine is grown for its flowers and glossy foliage. It is planted in small or large gardens for its fragrance or to grow over lattice or a pergola. It is suitable for coastal sub-tropical regions and establishes in 1–2 years. It is also used in containers and grown under glass in cold climates. Once established, it has a high-water requirement and responds to an occasional deep watering particularly during dry periods [10].

The production of jasmine is affected by various factors, among which insect pests are the most devastating factor. The major pests affecting jasmine are jasmine bud worm (*Hendecasis duplifascialis* Hampson), leaf webworm (*Nausinoe geometralis* Guenee), gallery worm (*Elasmopalpus jasminophagus* Hampson), leaf roller (*Glyphodes unionalis* Hübner), and the two spotted mite (*Tetranychus urticae* Koch). Of these, bud worm and blossom midge gain major economic importance, as they cause excessive damage to the buds, which is the economic part of the plant [11]. The midge maggots enter into the buds at the base of the corollas resulting in swelling and shrivelling at the base of the buds. The maggots feed inside unopened flower buds, causing deformed, pink discolored buds and blossoms. In case of severe infestations, buds dry prematurely leading to bud drop or blossom drop, thereby the marketable quantity of the flowers is greatly reduced [12] (**Figure 5**).

Management of this pest has become a menace for the jasmine growers. The knowledge on life history of the pest as well as the life table on different varieties is essential for developing IPM in better management of any pests. The knowledge of the sequence of developmental stages, their duration, and number of generations and method of overwintering is essential to know the 'weakest link' in the life cycle. This would help to aim control measures effectively at the most vulnerable stage of the pest. The biology of jasmine midge is attempted in the four cultivable jasmine

Figure 4.
Jasminum nitidum.

Figure 5.
Symptoms of damage of blossom midge, Contarinia maculipennis.

species, that is, *J. sambac*, *J. auriculatum*, *J. grandiflorum,* and *J. nitidum*. The objective of the study is to compare the different life stages of blossom midge in different *Jasminum* species.

6. Materials and methods

In the laboratory, 10 pairs of adult midges were introduced into the rearing cage (50 × 45 × 45 cm). A 250 ml conical flask filled with water plugged with adsorbent cotton and inserted with fresh *J. sambac* branches with flowers and buds was kept inside the cages to facilitate oviposition. The buds were checked for eggs, and the life history was studied. The lower surface of the rearing cage was filled with moist sand and planted with three-fourth mature flower buds to facilitate pupation of the maggots. The egg, maggot, pupal, and adult longevity periods were recorded. Jasmine branches with flowers were replaced daily with fresh branches.

The biology was also studied by rearing them in *J. auriculatum, J. grandiflorum,* and *J. nitidum* buds by placing the respective branches in the conical flask with flowers and buds (**Figure 6**).

To study the biology in different *Jasminum* species, the midges were cultured in *J. sambac, J. auriculatum, J. grandiflorum,* and *J. nitidum* placed inside Petri dishes, and the following parameters were recorded.

Figure 6.
Mass culturing of blossom midge.

7. Results and discussion

The life span of blossom midge, *C. maculipennis,* was studied on four cultivable *Jasminum* species namely, *J. sambac, J. auriculatum, J. grandiflorum,* and *J. nitidum* in the laboratory. The life cycle of blossom midge consisted of egg, larval, pupal, and adult stages (**Figure** 7). The extensive details of life stages were studied in all the four *Jasminum* sp. and furnished in **Table 1**. The different life stages were detailed in the following paragraphs.

7.1 Eggs

The eggs were elongated and cylindrically laid on the inner whorls of the petals in groups of 10–13 during night times. The eggs hatched in 1–2 days. The average egg period was 1.3, 1.2, 1.1, and 1.1 days in *J. sambac, J. auriculatum, J. grandiflorum,* and *J. nitidum,* respectively.

7.2 Maggot

The larval stages of midges are called maggots. There were totally four larval instars. The maggots were narrow with pointed anterior and posterior ends.

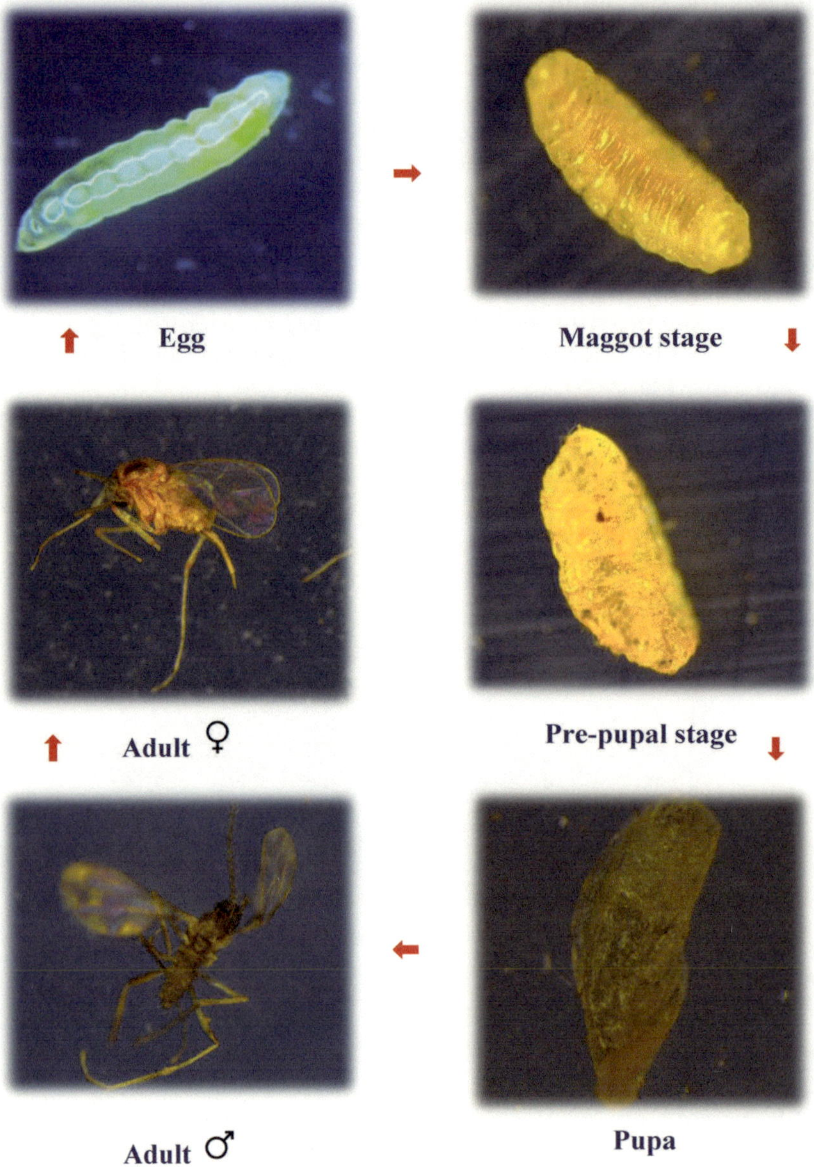

Figure 7.
Biology of blossom midge, Contarinia maculipennis.

The maggots on eclosion were dull white in color and turned yellow as the development progressed. The maggots scraped the petals, stamen, and stigma and suck sap from the flower buds. The larval period lasted for 4–5 days. The average larval period observed in *J. sambac, J. auriculatum*, *J. grandiflorum,* and *J. nitidum* was 4.4, 4.2, 4.1, and 3.9 days, respectively.

7.3 Pupa

The maggots turned orangish yellow before pupation. Pupation occurred in the top superficial layer of soil in a thin white case. The pupal period lasted for 7–8 days. The average pupal period was 7.5, 7.3, 7.2, and 7.0 days in *J. sambac, J. auriculatum*, *J. grandiflorum,* and *J. nitidum,* respectively.

Life stages of midge	Duration of different life stages of *C. maculipennis* in four *Jasminum* species, mean ± SD (in days)			
	J. sambac	*J. auriculatum*	*J. grandiflorum*	*J. nitidum*
Egg period	1.30 ± 0.483	1.20 ± 0.421	1.10 ± 0.31	1.10 ± 0.31
Larval period	4.40 ± 0.737	4.10 ± 0.42	4.10 ± 0.32	3.90 ± 0.32
Pupal period	7.50 ± 0.674	7.30 ± 0.52	7.20 ± 0.42	7.00 ± 0.00
Adult period	2.03 ± 0.874	1.90 ± 0.88	1.80 ± 0.79	1.90 ± 0.57
Total life cycle	15.20 ± 14.24	14.3 ± 1.004	13.3 ± 1.441	13.40 ± 1.07

Table 1.
Life stages of blossom midge, Contarinia maculipennis, *on four* Jasminum *species.*

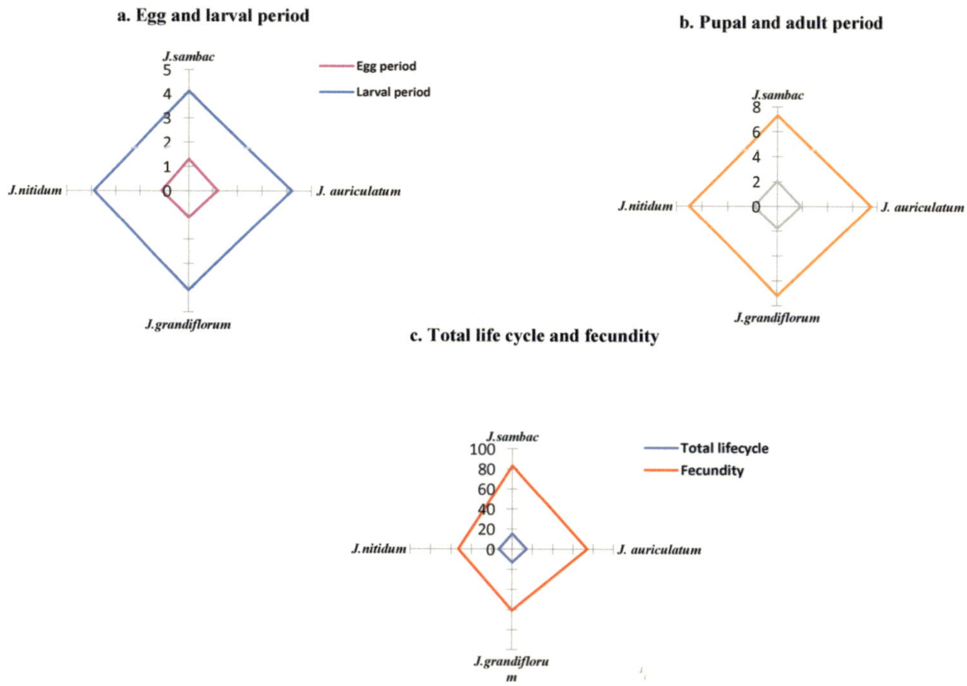

Figure 8.
Radar representation of biology of blossom midge, Contarinia maculipennis *Felt in different* Jasminum *species.*

7.4 Adult

The adults were minute and delicate flies. The females had a black head and yellowish brown body and were characterized by a distinct long ovipositor. The males were brownish and shorter than females. The moniliform antennal segments adorned with hairs in whorls were short and cylindrical in males and long and spherical in females. The adults usually lived for 1–2 days and for 3 days in rare cases. In an average, the adult period was observed to be 2.0, 1.9, 1.8, and 1.9 days in *J. sambac, J. auriculatum, J. grandiflorum,* and *J. nitidum,* respectively.

7.5 Total life span

The total duration taken to complete egg to adult stage was 15.4, 14.3, 13.3, and 13.4 days in *J. sambac, J. auriculatum, J. grandiflorum,* and *J. nitidum,* respectively (**Figure 8**).

8. Discussion

The average egg, maggot, pupal, and adult period were 1.3, 4.4, 7.3, and 2.0 days in *J. sambac*. The findings were in agreement with previous reports that cecidomyiid eggs hatched in 1.4 days, and the larval stage lasted 4–5 days, and adults emerged in 7–8 days [13]. Previous research works studied on the life stages of *C. maculipennis* and evolved the same results [14]. While comparing the biology in different *Jasminum* species, the egg period was highest (1.3 days) in *J. sambac* and least (1.1 days) in *J. grandiflorum* and *J. nitidum*. The trend was similar to other stages also. The total life cycle was highest (15.4 days) in *J. sambac* and least (13.4 days) in *J. nitidum* [15]. The longest duration of midges on *J. sambac* revealed the presence of physical factors like bud shape and size or chemical factors like high phenols, lower protein, and sugars that impair the faster growth and development. Host plants adversely affect the biology of the phytophagous insect, reducing the size, longevity, fecundity, and increasing mortality or alternatively may indicate that midges do not prefer to feed on *J. auriculatum, J. grandiflorum,* and *J. nitidum.*

9. Conclusion

The life cycle studies of blossom midge on four cultivable *Jasminum* sp., *J. sambac, J. auriculatum, J. grandiflorum,* and *J. nitidum* revealed that the life cycle consists of egg, maggot, pupal, and adult stages. The life cycle and duration of each stage are found to be the shortest on *J. nitidum* and the longest on *J. sambac.*

The use of moderately resistant varieties as a part of IPM strategy can enhance the biological and chemical tools of insect pest management. One of the techniques to decrease the pest damage is the use of the cultivars or species, which shows higher resistance to insect pests such as *J. nitidum* and *J. grandiflorum* in areas where incidence of jasmine mites was high. *J. grandiflorum* is a commercially cultivated species of Jasmine, well known for its mesmerizing fragrance as well as usage for concrete recovery. But, *J. nitidum* is a new *Jasminum* species with star-shaped fragrant flowers newly introduced for commercial flower cultivation. Hence, utilizing this least preferred species in breeding programs of jasmine will yield better varieties with greater degree of resistance to jasmine pests.

Author details

I. Merlin Kamala
Tamil Nadu Agricultural University, Coimbatore, India

*Address all correspondence to: merlinento@gmail.com

References

[1] Huxley AJ, editor. The New Royal Horticultural Society Dictionary of Gardening, 4 volumes. London, England: Macmillan Press; 1992. 3240 pp

[2] Wunderlin RP, Hansen BF. Guide to the Vascular Plants of Florida. 3rd ed. Gainesville, Florida: University Press of Florida; 2011. 783 pp

[3] Mabberley DJ. Mabberley's Plant-Book: A Portable Dictionary of Plants, Their Classification and Uses. 3rd ed. New York, New York: Cambridge University Press; 2008. 1021 pp

[4] Bhattacharjee SK. Native jasmine of India. Indian Perfumes. 1980;**24**(3):126-133

[5] Lakshmi J, Ganga M. Floral biology studies in certain lesser known species of jasmine (*Jasminum* spp.). International Journal of Current Microbiology and Applied Sciences. 2017;**6**(8):2811-2815

[6] Sharma AJR, Sharma AKS. *Jasminum* species: An overview. 2011. Available from: https://www.researchgate.net/publication/292575441_*Jasminum*_species_an_overview

[7] Kumari G, Wong KH, Serra A, Shin J, Yoon HS, Sze SK, et al. Molecular diversity and function of jasmintides from *Jasminum sambac*. BMC Plant Biology. 2018;**18**:144

[8] Manimaran P, Ganga M, Kannan M. Manipulation of petal senescence in *Jasminum nitidum* flowers with packaging and pre-treatment during storage: Role of phenolics. Journal of Pharmacognosy and Phytochemistry. 2018;7(2):3478-3480

[9] Sandeep, Paarakh PS. *Jasminum grandiflorum* Linn (Chameli): Ethnobotany, phytochemistry and pharmacology—A review. Pharmacology Online. 2009. Available from: https://www.researchgate.net/publication/215517134_*Jasminum*_*grandiflorum*_Linn_Chameli_Ethnobotany_Phytochemistry_and_Pharmacology_-_A_review [Accessed: 02 April 2019]

[10] Bhaguna Y, Singh RM, Juyal V, Gopt V. Antilithiatic effect of flowers of *Jasminum auriculatum* Vahl. International Journal of Green Pharmacy. 2009:654-672. DOI: 10.4103/0973-8258.54910

[11] Kamala IM. Incidence of blossom midge, *Contarinia maculipennis* Felt on jasmine (*Jasminum sambac* L.) and its natural enemies in Tamil Nadu. Indian Journal of Ecology. 2018;**45**(2):367-372

[12] Merlin Kamala I, Kennedy JS. Evaluation of entomopathogens against blossom midge, *Contarinia maculipennis* Felt in Jasmine (*Jasminum sambac* L.). Journal of Biological Control. 2018, 2018;**32**(2):121-127. DOI: 10.18311/jbc/2018/18923

[13] David PMM, Hanifa AM, Natarajan S. Evaluation of some insecticides and neem oil against jasmine blossom midge, *Contarinia* sp. Madras Agricultural Journal. 1991;**78**:82-84

[14] Neelima Y. Bioecology and management of jasmine pests [MSc (Ag) thesis]. Acharya N.G. Ranga Agricultural University; 2005. Submitted

[15] Merlin Kamala I. Bioecology and integrated management of major pests of Jasmine (*Jasminum sambac*) [PhD thesis]. Tamil Nadu Agricultural University; 2017. Submitted

Chapter 8

Phylogeny and Functional Morphology of Diptera (Flies)

Benjamin Kongyeli Badii

Abstract

The order Diptera includes all true flies. Members of this order are the most ecologically diverse and probably have a greater economic impact on humans than any other group of insects. The application of explicit methods of phylogenetic and morphological analysis has revealed weaknesses in the traditional classification of dipteran insects, but little progress has been made to achieve a robust, stable classification that reflects evolutionary relationships and morphological adaptations for a more precise understanding of their developmental biology and behavioral ecology. The current status of Diptera phylogenetics is reviewed in this chapter. Also, key aspects of the morphology of the different life stages of the flies, particularly characters useful for taxonomic purposes and for an understanding of the group's biology have been described with an emphasis on newer contributions and progress in understanding this important group of insects.

Keywords: Tephritoidea, Diptera flies, Nematocera, Brachycera metamorphosis, larva

1. Introduction

Phylogeny refers to the evolutionary history of a taxonomic group of organisms. Phylogeny is essential in understanding the biodiversity, genetics, evolution, and ecology among groups of organisms [1, 2]. Functional morphology involves the study of the relationships between the structure of an organism and the function of the various parts of an organism. The old adage "form follows function" is a guiding principle of functional morphology. It helps in understanding the ways in which body structures can be used to produce a wide variety of different behaviors, including moving, feeding, fighting, and reproducing. It thus, integrates concepts from physiology, evolution, anatomy and development, and synthesizes the diverse ways that biological and physical factors interact in the lives of organisms [3].

The order Diptera includes all true flies. These insects are distinctive because their hind wings are modified into small, club-shaped structures called halters. Diptera flies are the most ecologically diverse group of insects, and probably have a greater economic impact on humans than any other group of insects [4, 5]. Several families of Diptera are involved in the transmission of disease pathogens to humans and other animals. Biting flies cause annoyance that impacts negatively on tourism, recreation, land development, and industrial productivity. Some flies are pests of agricultural plants especially those that infest fruit crops in the field. On the other hand, many flies are beneficial; particularly those that pollinate flowering plants, assist in the decomposition of organic matter, or serve as biocontrol agents of insect pests [6].

Diptera is one of the most species-rich, anatomically varied and ecologically innovative orders of insects, making up 10–15% of known animal species. An estimated 195,000 species of Diptera have been described [7, 8], however, the actual total number of extant fly species is many times that number. The living dipteran species have been classified into about 17,000 genera, 190 families, 28–39 superfamilies and 12–15 infraorders, and around 3100 fossil species have been described [9]. The monophyly of Diptera is well established with a number of complex morphological modifications recognized as synapomorphies, including the transformation of the hindwings into halteres, and the development of the mouthpart elements for sponging liquids.

Recent research into the phylogeny of Diptera has been characterized by more sophisticated and consistent methods of analyzing traditional phylogenetic and morphological characters [10], and the inclusion of ever larger volumes of molecular sequence data. The greatest advances in dipteran phylogenetics over the past few decades have been made by a relatively small number of authors attempting to synthesize phylogenetic data across large components of it, by the use of quantitative methods. A clearer understanding of the bioecological processes of Diptera would necessitate a more robust estimate of their phylogenetics and functional morphology, which will serve as an organizing framework for fly classification and nomenclature and as the context for understanding the pattern of evolutionary change, tracing the origin of morphological and ecological adaptations, and documenting diversification itself. In this chapter, the current status of Diptera phylogenetics and functional morphology is reviewed with emphasis on newer contributions and progress in understanding these group of insects [11].

2. Phylogeny of Diptera

The monophyly of Diptera is generally accepted and exceedingly well supported, but there is still no consensus as to the resolution of this major insect group in higher level monophyletic units. Indeed, there are several very species-rich dipterous taxa which represent well corroborated monophyla: Brachycera, Cyclorrhapha (as Muscomorpha); and Schizophora. The phylogenetic relationships of these groups are presented here by considering the Diptera tree on a base-to-top approach.

2.1 The order Diptera

Diptera is the fourth insect order (after Coleoptera, Lepidoptera and Hymenoptera) in terms of number of known and described species [12]. The order contains about 195,000 named species in approximately 190 families, with thousands of species of agricultural, medical and veterinary importance [13]. Diptera is considered to be the most ecologically diverse order of the Insecta class [14]. The dietary composition of Diptera flies basically consists of a wide range of hemophagy and parasitism (both endo- and ecto-parasitism) of vertebrates and invertebrates, and several forms of mycetophages, saprophages and phytophages [15]. Most species of this order are known to be present in almost all zoogeographic areas of the world. They have proved to be well adaptable to a wide diversity of habitats; they can survive in almost every habitable environment on the earth, except the ocean depths [16].

Dipterans are holometabolous endopterygote insects, undergoing complete metamorphosis, in which a pupal stage intervenes between the larval and adult instars. Immature stages are morphologically different from adult forms, and often have contrasting habitat and food requirements [17]. Although some families, species, and sometimes members of one sex of flies are apterous (possess no wings), Diptera as a whole can be characterized for possessing only two functional front wings, and a pair of vestigial knob (named halters) behind the wings, that function as organs of equilibrium, helping the flies to remain stable during flight (**Figure 1**).

Two suborders can be recognized in Diptera; Nematocera and Brachycera (**Figure 2**). The Nematocerans include cane flies, midges, mosquitoes and gnats, which have thin multisegmented antennal flagella. Larval forms generally have conspicuous head, and pass through more than three instars before reaching the pupal stage. The Brachycerans include higher flies such as hover flies and dung flies, which possess shorter and thicker antennae with fewer flagellomer. These have robust bodies with legs shorter than the Nematocerans. Brachyceran larval forms have the posterior portion of the head capsule desclerotized and extended into the thorax.

The phylogenic relationships between these two suborders are still controversial, although Nematocera is suspected to be paraphyletic [18]. The suborder Brachycera consists of the infraorder Muscomorpha (=Cyclorrapha) which encompasses species with larval forms commonly referred to as maggots. These larval forms are dominantly saprophagous, and are morphologically soft bodied with mouthparts modified into sclerotized hooks for feeding into host substrates. In almost all Cyclorraphan flies, pupation is internal, occurring within the puparium (a tanned cuticle covering the pupa) during the final larval instar stage.

Within the Cyclorrapha, the division Schizophora comprised the largest tertiary radiation of insects, with approximately 50,000 species. The suborder Schizophora has distinctive morphological features such as the balloon-like membraneous sac which usually ruptures the puparium during the emergence of the adult. This balloon-like structure is referred to as the ptilinum, which may invaginate into

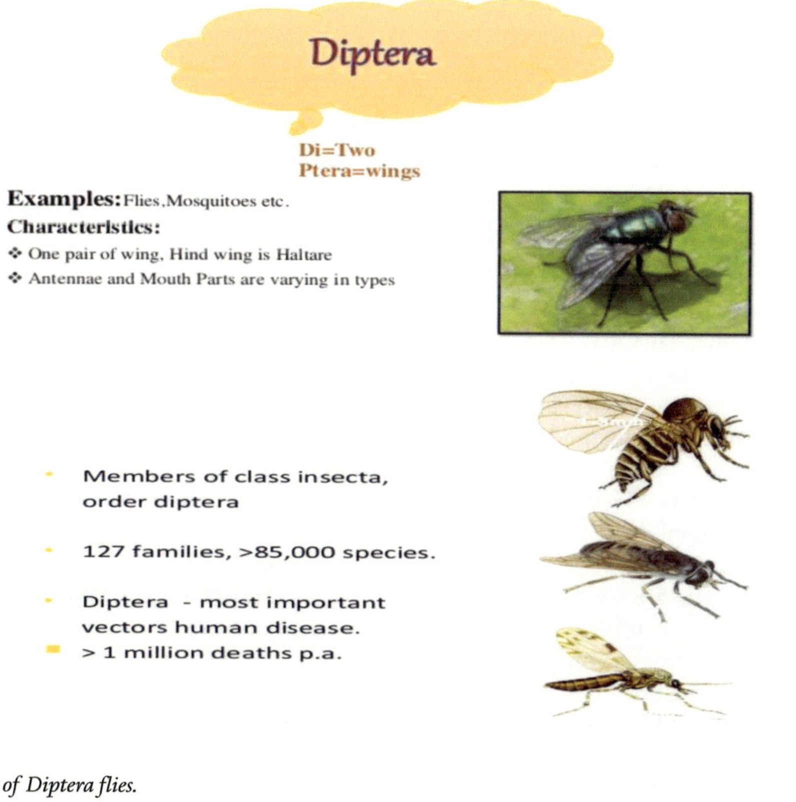

Figure 1.
Key features of Diptera flies.

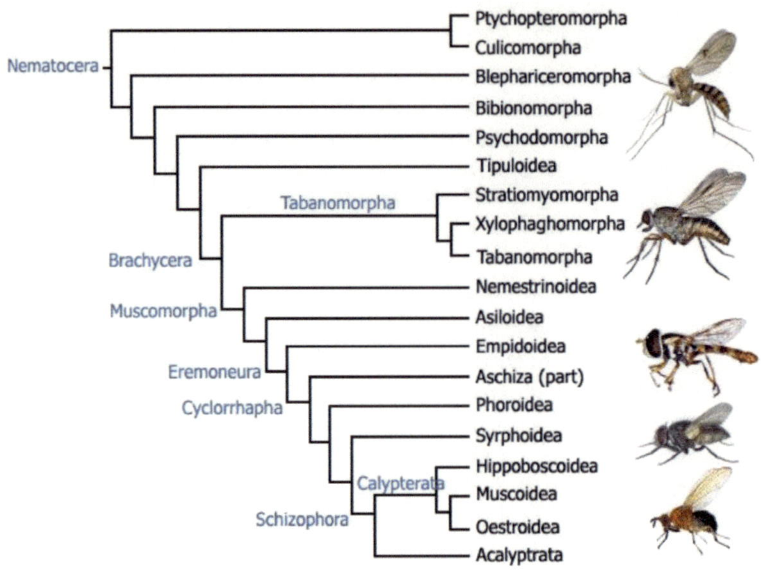

Figure 2.
Phylogenic tree of Diptera.

the capsule of the head, resulting in the face of the adult being bordered by a form ptilinal fissure. The section Acalyptratae is usually found within the Schizophora suborder especially species of the group that do not possess any calypters (lobe-like structures at the extreme basal part of the wings). Members of Acalyptratae belong to families in the superfamily Tephritoidea.

2.2 The superfamily Tephritoidea

The superfamily Tephritoidea includes eight known families of acalypterate flies: Lonchaeidae, Piophilidae, Pallopteridae, Richardiidae, Ulidiidae, Plastytomatidae, Pyrgotidae, and Tephritidae [19] (**Figure 3**). Lonchaeidae are commonly known as lance flies. There are about 500 described species in 9 genera. Lonchaeids are generally small, robustly built flies with blue-black or metallic bodies. Lonchaeid larvae are secondary invaders of diseased bodies or injured plant tissues, adults have the rare habit among acalypterate flies to form swarms for mating, and are found mostly in wooded areas.

The Piophilidae mostly includes scavengers in fungi and animal products, with the family getting its common name, skipper flies, due to the fact that larvae tend to jump during their last instars before pupation. Pallopteridae or flutter-winged flies, is a little-known small family, with larvae of some species feeding in flower buds, or occurring as predators of wood borer larvae under the bark of fallen trees. Over 50 species in 15 genera are found in the temperate region of the Northern and Southern hemispheres. The Richardiidae is a small family that consists of 30 genera and 175 species. It is a little-known family whose adults can be captured in fruit-baited traps; the few larval feeding records of this family suggest that these flies feed on rotten vegetable matter. Most adults generally have conspicuously pictured wings, often with metallic-blue or greenish colors on bodies and legs, and a typical tephritoid ovipositor.

Ulidiidae and Plastytomatidae are both pictured-winged flies. The Plastytomatids are sometimes referred to as signal flies. Both families are abundant in the tropics, occurring in decaying tissues but also sometimes feeding on plants

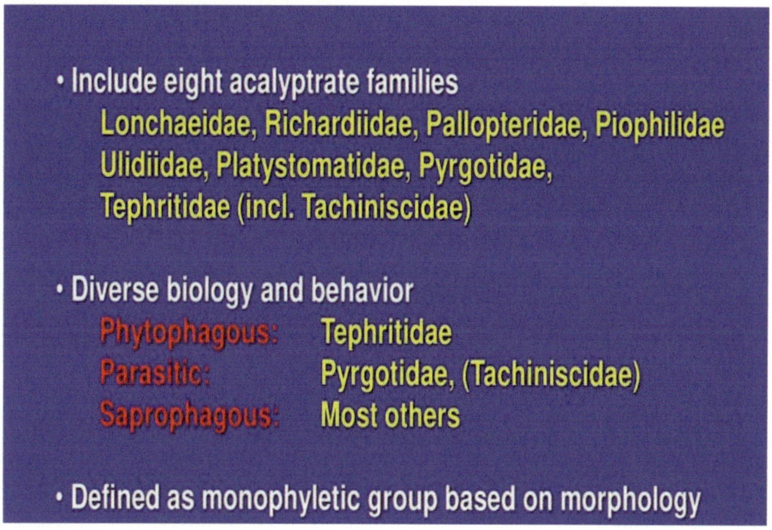

Figure 3.
Composition of the superfamily Tephritoidea.

with a few species considered as pests. Most species share with the Tephritidae by an unusual elongated posteroapical projection of the anal cell in the wing, but can be differentiated by the smooth-curving subcostal veins. Pyrgotidae comprises medium to large flies with considerable coloring of the wings. They are mostly nocturnal in habit. Unlike other tephritoids, Pyrgotids are endoparasitoids; the females pursue scarab beetles in flight, laying their eggs on the back of the beetle, under the elytra and beyond host's reach. Developing larvae enter the cavity of the beetle and eventually kill the host before pupation. Phylogenetic relationships among tephritoid families have been reviewed by previous authors.

2.3 The family Tephritidae

Tephritidae (true flies) is a very large family, which includes more than 4000 described species. Worldwide members of this family are among the most economically crucial insect pests of edible fruits and vegetables [20, 21]. The family can be characterized by an elaborate wing patterns and the possession of a telescopic ovipositor by the female. Tephritidae is known as one of the most ecologically diverse families of Diptera, and due to its size, it has been difficult to synthesize phylogenetic relationships among higher groups of the family [22]. Phylogenetic relationships of important genera of the family have been provided by [23]. Despite the lack of a conclusive phylogeny, the study of Tephritidae can be approached by looking separately at five different subfamilies; Blepharoneurinae, Phytalmiinae, Trypetinae, Dacinae and Tephritinae, all of which are well represented in the tropics.

The subfamily Blepharoneurinae represents flies of the tropical group, and composed of five main genera; *Ceratodacus, Problepharoneura, Blepharoneura, Baryglossa,* and *Hexaptilona.* The first three genera consist of species of the neotropical and afrotropical regions, while the last two genera include species of the palearctic regions. Although this subfamily is composed of a reduced number of described species, recent scrutiny on flies in the genus *Blepharoneura* suggests that there may be more than 200 species. This subfamily is interesting as the group appears to be one of the oldest lineages in Tephritidae. All the genera for which biological data have been gathered feed on plants and parts of plants in the family Cucurbitaceae. There is suggestive evidence that these flies have undergone rapid processes of speciation, as much as more than one species can cohabit the same plant, exploiting different parts of it and exhibiting complex courtship behaviors [24].

Phytalmiinae is a subfamily comprising six genera; *Diplochorda, Ortaloptera, Phytalmia, Sessilinia, Tetrastiomyia,* and *Sophiria.* These are the flies with antenna-like head projections and sometimes referred to as antler flies or deer flies (not to be confused with Tabanidae). Decaying plant material is the larval food across this subfamily. All described species of antler flies occur between the island of Borneo and the Cape York Peninsula of Australia. The few behavioral studies on this group suggest that antler flies have been evolved in the context of male intrasexual competition. Resource defense mating systems for this group have been described by [25], while [26] provided a review on current knowledge on the Phytalmiinae.

Trypetinae is a large subfamily that includes 19 known genera; *Carpomya, Cryptodacus, Goniglossum, Haywardina, Myiopardalis, Rhagoletis, Rhagoletotrypeta, Zonosemata, Acidia, Euleia, Strauzia, Trypeta, Anastrepha, Toxotrypana, Epochra, Paraterellia, Chetostoma, Oedicarena,* and *Myoleja*. The genera *Rhagoletis, Anastrepha,* and *Toxotrypana* include several species of major economic importance. While members of *Rhagoletis* are both holarctic and neotropical in distribution, *Anastrepha,* and *Toxotrypana* are restricted to the new world, while the rest of Trypetinae is especially diverse in the old-world tropics, which may be the center of origin [27]. Within Trypetinae, the subtribe Trypetina contains all the known leaf-mining species of tephritids, along with others by different larval feeding habits. A comprehensive account of this group of flies is provided by [28].

Tephritinae is considered the most specialized subfamily of Tephritidae. It is composed of six tribes with over 210 genera. Most species of Trypetini breed in flower heads, or form flower, stem, or root galls in plants of the family Asteraceae. Due to this habit, many of these tephritids have been used in biological control of weeds [29, 30]. Sexual behavior and biology of some members of this subfamily have been reviewed by previous authors [31].

Dacinae is a subfamily that contains only three genera *Bactrocera, Dacus,* and *Ceratitis* all of which include many species of major economic importance. All members of this subfamily are native to the Old World, despite the fact that the Mediterranean fruit fly, *C. capitata* has been established in South and Central America since the beginning of the 20th century, and there have been recurrent introductions and eradication efforts of this pest along with the Olive fruit fly, *B. oleae,* the Oriental fruit fly, *B. dorsalis* and others in North America.

2.4 True fruit flies

The term "fruit fly" is sometimes used for two distantly related groups of flies, namely the families Drosophilidae and Tephritidae (**Figure 4**). The Drosophilidae includes "fruit flies" of the geneticists, which are in reality, micro-fungi feeders that have acquired this name because of their habit of feeding on decaying fruit [32]. The Tephritidae is generally described to include the "true fruit flies" because most species attack living plant material, and an estimated 40% of the over 5000 described species attack intact and growing fruits. Females of fruit flies have an ovipositor, similar to the "sting" of a wasp, with which they puncture the skin of healthy fruits and lay their eggs therein. Larval development is completed within the fruit (which may become rotten as a result) and the fully-grown larvae then drop into the soil and form a puparium.

There are about 150 genera and 950 species of Tephritid fruit flies known in tropical Africa, most of which form a natural component of Africa's rich and varied

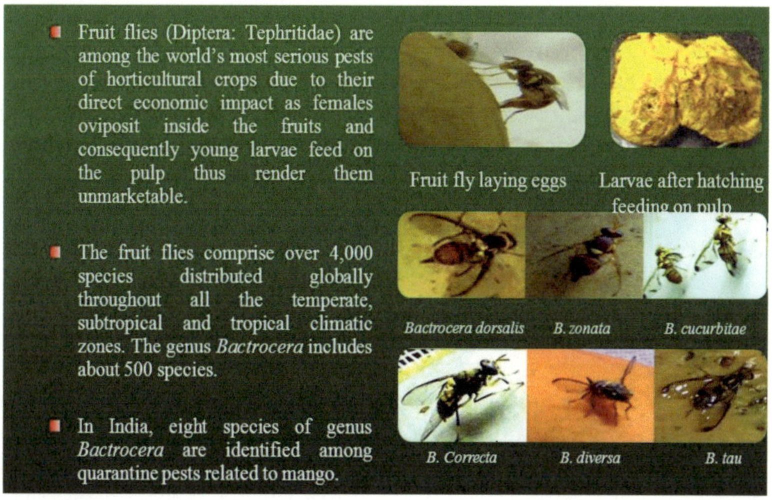

Figure 4.
True fruit flies (Diptera: Tephritidae).

biodiversity. About 70 species of fruit flies are considered important agricultural pests, and many others are minor or potential pests. Fruits and vegetables are the most important crops attacked [33–35], even though some seed crops are also affected.

3. Functional morphology of Diptera

Morphology and anatomy of the different life stages of fruit flies, particularly their characters useful for taxonomic purposes, have been described in detail by previous authors. But only those aspects relevant for an understanding of the group's developmental biology are considered here.

3.1 The adult

Adults of Diptera have segmented bodies that include a head, thorax, and abdomen. The head has large eyes that cover the sides of the head with a small space between them in the front of the head. This helps them to see a wider area as they are flying (**Figure 5**). The adult body coloration of different dipteran species however varies from black through various shades of brown to orange or yellow. Yellow marks, particularly on the thorax, give many species a somewhat wasp-like appearance. This resemblance is particularly pronounced in certain *Bactrocera* subgenera and *Callantra* spp., which have petiolate abdomens, heavily fuscated costal stripes on the wings, and a jerky, wasp-like walk. The paired antennae each consist of three segments. Scanning electron microscope studies on *B. oleae* and *B. tryoni* indicate that the outer segment is covered with long cuticular spines interspersed with large numbers of chemosensilla of several distinct morphological types and functional significance [36–38].

The general structure of dacine flies is fairly typical of cyclorrhaphan Diptera. Male dacines, except those of some groups such as Gymnodacus, typically have a pair of combs (or pectins) comprised of stiff curved bristles on the lateral hind margins of the third abdominal tergite.

These combs function as stridulatory organs during courtship. Both sexes have a pair of tergal glands (ceromae) that open onto the surface of the fifth tergite. These consist of dense groups of minute alveolae that secrete a waxy substance, which is spread onto the body and wings during preening [39]. In female dacines, abdominal segments 7–9 form the ovipositor, which is usually smooth and pointed but is serrated in some species. The apical segment has a number of chemosensilla; the most prominent are the preapical setae that arise from lateral grooves on either side of the segment [40]. These presumably play an important role in fruit discrimination.

3.2 The egg

The study of dipteran egg morphology has largely been confined to the description of surface features [41] mainly through transmission electron microscopy examination [42]. Typically, eggs of Diptera flies are elongate ellipsoidal in shape and thus have only a single primary axis (**Figure 6**). At one end, the egg bears a pedicel. The pedicel bears the micropyle and the aeropyles. Typically, the micropyle is located on the apex of the pedicel and may have a single- or multiple-openings. Among the majority of Diptera flies, the shape of the egg is usually elliptic or ovoid with elongated appearance. The end portion may be blunt, round or fusiform, subglobose with about 1 mm in size. The length-width ratio may vary depending on the species. In several species of Sarcophagidae, however, the egg length may be 2.5 cm or longer [43]. Egg coloration widely varies within this order, and may range from pale (shortly after oviposition) to dark (towards embryo development). The micropyle may be arranged in a similar manner as in most nonfrugivorous species. The pedicel may be only a slight outgrowth or an elongated stalk nearly as long or longer than the overall length of the egg [44].

Dipteran eggs usually develop inside the ovariole during which the pedicel begins to orient towards the terminal portion of the ovary. It is understood that the basic functions of fertilization and oviposition are facilitated by this orientation process. As observed by [45], fertilization usually occurs as the egg moves

Figure 5.
Dorsal view of adult of Dacine fly.

Figure 6.
Matured eggs of Tephritid fly exposed from an infested fruit.

towards the middle portion of oviduct through the micropyle. The gonopore exists within the basal part, near the end of the aculeus (the part inserted into the tissues of the plant during oviposition). Shortly after oviposition, the process of embryogenesis commences. During the process of embryogenesis, the developing head of the embryo then begins its orientation towards the pedicel. In many species, however, just before eclosion, the embryo is said to rotate 180° and then leaves the egg through the basal end. This process makes it possible for the embryo to be positioned so that the plant tissue can be easily encountered upon eclosion from the egg [46].

The surface of a matured egg of Diptera may appear smooth or rough (due to the presence of microsculptures containing chorion derivatives) Also, the egg surface may have polygonal reticulations (mass-relief-type ridges) which represent the follicle cells outlines and are responsible for the lay down of the chorion [47]. For *Tephritis baccharis* (Coquilett), the reticulations may appear more prominent and may bear further structural decorations. For *Aciurina thoracica* (Curran), the egg surface may develop as rough at the end of the pedicel, and then diminish to a smooth surface close to the basal end. Previous studies hypothesized that the pedicel usually needs greater structural support for protecting the aeropyles and associated channels of respiration from deformation. This is because the pedicel portion of the egg is exposed to facilitate exchange of gases, while the basal end is inserted into tissues of the host plant [48].

3.3 The larvae

Larvae are the small wormlike early stages of Diptera flies, usually called maggots. Larvae of lower Diptera range in length from only a few millimeters to many centimeters, depending on the species, and are usually distinguished by having a conspicuous head capsule with opposable mandibles that move in a pincer-like horizontal plane. After eggs hatch, larvae begin to feed on the decaying materials within which they were laid (**Figure 7**). Larvae consume as much food as possible in order to store energy and nutrients for the upcoming pupal stage. Three free-living instars exist for tephritid fruit flies. The only known exceptions are *Urophora jaceana* (Herring) and *Urophora cardui* (Linnaeus), in which the first instar remains in the egg and exits as a second instar. The external anatomy of the larvae of frugivorous tephritids has been examined in detail, and at least partial descriptions based primarily on scanning electron micrographs for 25 species have been available.

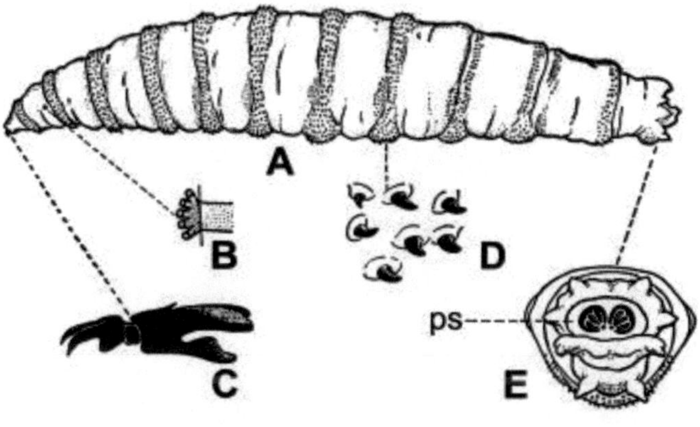

Figure 7.
Blowfly larva, Chrysomya bezziana *(Calliphoridae). (A) Complete larva; (B) anterior spiracle; (C) cephalopharyngeal skeleton; (D) spines; (E) caudal end with pair of spiracular plates.*

By comparison, an atlas of immature morphology based on the third instar of 34 economically important species has been developed. The segments of the maggot typically bear spines in regular patterns (**Figure 7**) and the larvae of some species may possess structures that vary from simple setae to large protuberances. Several other structures including the median oral lobe, the lateral spiracles accompanied by a variable number of sensilla associated with the sensory organs of the gnathocephalon have been newly identified for various frugivorous species.

Tephritids have distinct anterior and posterior spiracles. Modern scanning electro-microscopy have aided in the location of the lateral spiracles along the thoracic and abdominal segments, as well as along the caudal segment that bears the posterior spiracles [49]. The lateral spiracles, which are always present along the lateral and anterior portion of a segment, have been known to have a varying number of campaniform sensilla associated with it around the posterior end of the spiracle. The number of sensilla may range from one (as in some *Aciurina* and *Trupanea* species) to as many as four (as in *Stenopa affinis* Quisenberry). When more than one spiracular sensilla is present, they are typically arranged adjacent to the spiracle, along a dorso-ventral axis.

3.4 The pupa

Dipteran pupa becomes more impervious to the surrounding environmental conditions and the larva becomes morphologically reduced and evolved to feed on nutrient-rich substrates; flies as a whole may occupy a broad range of trophic niches. The puparium is the hardened, penultimate larval integument of the developing fly (**Figure 8**). It is remarkable in its external morphology in tephritid fruit flies. When the third instar larva is ready to pupate, it leaves the medium, and its anterior spiracles evert, body shortens and ceases to move and it attaches to a firm substrate. The cuticle then transforms into a puparium, which is initially soft and white, but soon hardens, turning tan and eventually becomes brown and with bristle. Shortly after the puparium forms, then metamorphosis then takes place. The prepupal integument is shed and adheres to the inner wall of the puparium. The pupa forms within the puparium after the prepupal molt. The pupa develops independently of the puparium and has bilobed thoracic spiracles for respiration. The larval tracheae adjacent to the anterior and posterior spiracular openings remain open, thus allowing for gas exchange for the developing pupa within the puparium.

According to Ref. [50], there exists a pre-puparial stage in which the mouthparts contain series of invaginations. During this stage, the integument typically assumes a waxy appearance, but the processes of tagmentation (hardening and darkening of the integument) are delayed. The darkening of the integument may be triggered

Figure 8.
Pupal forms of Diptera.

by changes in the prevailing environment, by overwintering prepuparia as in the case of certain *Neaspilota* and *Urophora* species, particularly those found at higher altitudes [51].

Eclosion marks the end of pupation and the beginning of the adult life. The insect cracks open the puparium anteriorly and laterally at its seams and emerges from the pupal case. This almost invariably occurs around dawn, when leaves are still dump with dew, and the emerging fly can fold its new wings and harden its cuticle without the risk of desiccation. The timing of this is controlled by circadian rhythm.

4. Conclusion

These phylogenetic and morphological reviews of Diptera provide an evolutionary framework for future comparative work on species that are critically important to both society and science. The order Diptera has been divided into two or three suborders: the monophyletic Nematocera and the paraphyletic Brachycera, with the latter being divided further into the Orthorrhapha and Cyclorrhapha. The living dipteran species have been classified into about 10,000 genera, 150 families, 22–32 superfamilies and 8–10 infraorders. The typical dipteran body morphology is reflected in its life cycle which includes a series of distinct stages or instars; consisting of a brief egg stage, three or four instars, a pupal stage of varying length, and an adult stage that lasts from less than 2 hours to several weeks or even months. The species-richness, morphological variability and ecological diversity of this order of insects dictate the economic importance of the group to man and reflects the range of organisms in the order. Future work will focus on contributions and progress in understanding of the bioecological processes and economic impacts of dipteran flies in human life especially in relation to health, agricultural productivity and food security.

Conflict of interest

Author has no conflict of interest.

Author details

Benjamin Kongyeli Badii
University for Development Studies, Tamale, Ghana

*Address all correspondence to: benbadii@uds.edu.gh

References

[1] Amorim DS. A new phylogeny and phylogenetic classification for the Canthyloscelidae (Diptera: Psychodomorpha). Canadian Journal of Zoology. 2000;**78**:1067-1077

[2] Bertone MA, Courtney GW, Wiegmann BM. Phylogenetics and a timescale for diversification of the earliest true flies (Insecta: Diptera) based on multiple nuclear genes. Systematic Entomology. 2008;**33**:668-687

[3] Borror DJ, Triplehorn CA, Johnson NF. An Introduction to the Study of Insects. Fort Worth, Texas, U.S.A.: Saunders College Publishing; 1992. p. 875

[4] Pape T. Phylogeny and evolution of the bot flies. In: Colwell D, Scholl P, Hall M, editors. The Oestrid Flies: Biology, Host-Parasite Relationships, Impact and Management. Wallingford and Cambridge, MA: CABI Publishers; 2006. pp. 20-50

[5] Badii KB, Billah MK, Afreh-Nuamah K, Obeng-Ofori D. Review of the pest status, economic impact and management of fruit-infesting flies (Diptera: Tephritidae) in Africa. African Journal of Agricultural Research. 2015;**10**(12):1488-1498

[6] Merritt RW, Courtney GW, Keiper JB. Diptera (Flies, Mosquitoes, Midges, Gnats). In: Resh VH, Cardé RT, editors. Encyclopedia of Insects. San Diego CA, USA: Academic Press; 2003. p. 340

[7] Thompson FC. Biosystematic Database of World Diptera. Version 7.5. 2005. p. 409. Available from: http://www.diptera.org/biosys.htm [Accessed: 03 January 2018]

[8] Yeates D, Wiegmann DK. Phylogeny of diptera. Suricata. 2017;**4**:253-265. Available from: https://www.researchgate.net/publication/322220478 [Accessed: 03 January 2018]

[9] Yeates DK, Wiegmann BM, Courtney GW, Meier R, Lambkin C, Pape T. Phylogeny and systematics of Diptera: Two decades of progress and prospects. In: Zhang ZQ, Shear WA (eds.) Linnaeus Tercentenary: Progress in Invertebrate Taxonomy. Zootaxa. 2007;**1668**:565-590

[10] Sinclair BJ, Cumming JM. The morphology, higher level phylogeny and classification of the Empidoidea (Diptera). Zootaxa. 2006;**1180**:1-172

[11] Woodley NE, Borkent A, Wheeler TA. Phylogeny of the Diptera. In: Brown BV, Borkent A, Cumming JM, Wood DM, Woodley NE, Zumbado MA, editors. Manual of Central American Diptera. Vol. 1. Ottawa: NRC Research Press; 2009. pp. 79-94

[12] Badii KB. Knowledge Gaps, Training Needs and Bioecological Espects of Fruit-Infesting Flies (Diptera: Tephritidae) in Northern Ghana. PhD Thesis. Legon, Ghana: University of Ghana; 2014. p. 242

[13] Diniz IR, Morais HC. Classification and ecology of major tropical insect groups. In: Kleber H, Del C, Paulo S, Rico-Gray OV, editors. Tropical Biology and Conservation Management. Oxford, UK: UNESCO, EOLSS Publishers; 2008. pp. 202-213

[14] Mendes J. Diptera of tropical savannas. In: Kleber D, Claro C, Paulo S, Oliveira VRC, editors. Tropical Biology and Conservation Management, Encyclopedia of Life Support Systems (EOLSS), Developed under the Auspices of the UNESCO. Oxford, UK: Eolss Publishers; 2008. Available from: http://www.eolss.net [Accessed: 30 August 2012]

[15] Rull J. Phylogeny, biology, behavior, and management of tephritid fruit flies: An overview. In: Kleber D, Claro C, Paulo S, Oliveira VRC, editors. Tropical Biology and Conservation Management, Encyclopedia of Life Support Systems (EOLSS), Developed under the Auspices of the UNESCO. Oxford, UK: Eolss Publishers; 2008. Available from: http://www.eoss.net [Accessed: 30 August 2012]

[16] Grimaldi D, Engel MS. Evolution of the Insects. Hong Kong: Cambridge University Press; 2005. p. 755

[17] Romoser WS, Stoffolano JG. The Science of Entomology. Dubuque, Iowa, U.S.A.: W. C. Brown Publishers; 1994. p. 532

[18] Mittelbach GG, Schemske DW, Cornell HV, Allen AP, Brown JM, Bush MV, et al. Evolution and the latitudinal diversity gradient: Speciation, extinction and biogeography. Ecology Letters. 2007;**10**:315-331

[19] Korneyev VA. Phylogenetic relationships among the families of the superfamily Tephritoidea. In: Aluja M, Norrbom AL, editors. Fruit Flies (Diptera: Tephritidae): Phylogeny and Evolution of Behavior. Boca Raton, Florida: CRC Press; 2000. pp. 3-22

[20] Sarwar M. Area-wide integrated Management of Fruit Flies (Diptera: Tephritidae) pests in vegetables cultivation. Journal of Biological and Environmental Engineering. 2016;**1**(2):10-16

[21] Sarwar M. An area-wide integrated Management of Fruit fly (Diptera: Tephritidae) pests in fruits production. International Journal of Plant Science and Ecology. 2018;**4**(1):1-7

[22] Korneyev VA. Phylogenetic relationships among higher groups of Tephritidae. In: Aluja M, Norrbom AL, editors. Fruit Flies (Diptera: Tephritidae): Phylogeny and Evolution of Behavior. Boca Raton, Florida: CRC Press; 2000. pp. 73-114

[23] Norrbom AL, Thompson FC. Richard Herbert Foote (1918-2002)—Obituary. Proceedings of the Entomological Society of Washington. 2003;**105**:508-516

[24] Condon MA, Norrbom AL. Three sympatric species of *Blepharoneura* (Diptera: Tephritidae) on a single species of host (*Gurania spinulosa* (Cucurbitaceae)): New species and new taxonomic methods. Systematic Entomology. 1994;**19**:279-304

[25] Dodson GN. Resource defense mating system in antlered flies, Phytalmia spp. (Diptera: Tephritidae). Annals of the Entomological Society of America. 1997;**90**:496-504

[26] Dodson GN. Behavior of the Phytalmiinae and evolution of antlers in tephritid flies. In: Aluja M, Norrbom AL, editors. Fruit Flies (Diptera: Tephritidae): Phylogeny and Evolution of Behavior. Boca Raton, Florida: CRC Press; 2000. pp. 175-184

[27] Foote RH, Blanc FL, Norrbom AL. Handbook of the Fruit Flies (Diptera: Tephritidae) of America North of Mexico. Ithaca: Comstock Publishing Associates; 1993. p. 571

[28] Han HY. Phylogeny and behavior of flies in the tribe Trypetini (Trypetinae). In: Aluja M, Norrbom AL, editors. Fruit Flies (Diptera: Tephritidae): Phylogeny and Evolution of Behavior, CRC Press. Boca Raton: Florida; 2000. p. 297

[29] White IM, Elson-Harris MM. Fruit Flies of Economic Significance: Their Identification and Bionomics. Wallingford, U.K.: CAB Int; 1992. p. 601

[30] Turner CE. Tephritid flies in the biological control of yellow starthistle. In: McPheron BA, Steck GJ, editors.

Fruit Fly Pests: A World Assessment of their Biology and Management. Delray Beach: St. Lucie Press; 1996. pp. 171-176

[31] Headrick DH, Goeden RD. Behavior of flies in the subfamily Tephritinae. In: Aluja M, Norrbom AL, editors. *Fruit Flies (Diptera: Tephritidae):* Phylogeny and Evolution of Behavior. Boca Raton, Florida: CRC Press; 2000. pp. 671-707

[32] Aluja M, Rull J, Sivinski J, Norrbom AL, Wharton RA, Macias-Ordoñez R, et al. Fruit flies of the genus *Anastrepha* (Diptera: Tephritidae) and associated native parasitoids (hymenoptera) in the tropical rainforest biosphere reserve of Montes Azules, Chiapas, Mexico. Environmental Entomology. 2003;**32**:1377-1385

[33] Sarwar M. Occurrence of insect pests on guava (*Psidium guajava*) tree. Pakistan Journal of Zoology. 2006;**38**(3):197-200

[34] Sarwar M, Hamed M, Yousaf M, Hussain M. Monitoring of population density and fruit infestation intensity of Tephritid fruit flies (Diptera: Tephritidae) in *Citrus reticulata* Blanco orchard. Journal of Zoological Sciences. 2014;**2**(3):1-5

[35] Badii KB, Billah MK, Afreh-Nuamah K, Obeng-Ofori D. Species composition and host range of fruit-infesting flies in northern Ghana. International Journal of Tropical Insect Science. 2015;**35**:137-151

[36] Drew RAI, Courtice AC, Teakle DS. Bacteria as a natural source of food for adult fruit flies (Diptera: Tephritidae). Oecologia. 1983;**60**:279-284

[37] Drew RAI, Zalucki MP, Hooper GHS. Ecological studies of eastern Australian fruit flies (Diptera: Tephritidae) in their endemic habitat. I. Temporal variation in abundance. Oecologia. 1984;**64**:267-272

[38] Giannakakis A, Fletcher BS. Morphology and distribution of antennal sensilla of *Dacus tryoni* (Froggatt) (Diptera: Tephritidae). Journal of the Australian Entomological Society. 1985;**24**:31-35

[39] Munro HK. A taxonomic treatise on the Dacidae (Tephritoidea, Diptera) of Africa. Entomology Meseum of South African Department of Agriculture. 1994;**61**:1-313

[40] Hardy DE. Taxonomy and distribution of the oriental fruit fly and related species (Tephritidae-Diptera). Proceedings of the Hawaiian Entomological Society. 1969;**20**:395-428

[41] Headrick DH, Goeden RD. Life history and description of immature stages of *Aciurina thoracica* (Diptera: Tephritidae) on *Baccharis sarothroides* in southern California. Annals of the Entomological Society of America. 1993;**86**:68-79

[42] Margaritis LH. Comparative study of the eggshell of the fruit flies *Dacus oleae* and *Ceratitis capitata* (Diptera: Trypetidae). Canadian Journal of Zoology. 1985;**63**:2194-2206

[43] Knio KM, Goeden RD, Headrick DH. Comparative biologies of the cryptic, sympatric species, *Trupanea bisetosa* and *T. nigricornis* (Diptera: Tephritidae) in southern California. Annals of the Entomological Society of America. 1996;**89**:252-260

[44] Goeden RD, Teerink JA. Life history and descriptions of adults and immature stages of *Aciurina semilucida* (bates) (Diptera: Tephritidae) on *Chrysothamnus viscidiflorus* (hooker) Nuttall in southern California. Proceedings of the Entomological Society of Washington. 1996;**95**:59-78

[45] Headrick DH, Goeden RD. Reproductive behavior of California fruit flies and the classification and

evolution of Tephritidae (Diptera) mating systems. Studies in Dipterology. 1994;**1**:195-252

[46] Headrick DH, Goeden RD. Life history of *Trupanea californica* Malloch (Diptera: Tephritidae) on *Gnaphalium* spp. in southern California. Proceedings of the Entomological Society of Washington. 1991;**93**:559-570

[47] Mouzaki DG, Margaritis LH. Choriogenesis in the medfly *Ceratitis capitata* (Wiedemann) (Diptera: Tephritidae). International Journal of Insect Morphology and Embryology. 1991;**20**:51-68

[48] Goeden RD, Headrick DH. Life history and descriptions of immature stages of *Tephritis baccharis* (Coquillett) on *Baccharis salicifolia* (Ruiz and Pavon) persoon in southern California (Diptera: Tephritidae). Pan-Pacific. Entomology. 1991;**67**:86-98

[49] Headrick DH, Goeden RD. Resource utilization by larvae of *Paracantha gentilis* (Diptera: Tephritidae) in capitula of *Cirsium californicum* and *C. proteanum* (Asteraceae) in southern California. Proceedings of the Entomological Society of Washington. 1990;**92**:512-520

[50] Goeden RD, Headrick DH. Life history and descriptions of immature stages of *Neaspilota viridescens* Quisenberry (Diptera: Tephritidae) on native Asteraceae in southern California. Proceedings of the Entomological Society of Washington. 1992;**94**:59-77

[51] Goeden RD, Headrick DH, Teerink JA. Life history and description of immature stages of *Urophora timberlakei* Blanc and Foote (Diptera: Tephritidae) on native Asteraceae in southern California. Proceedings of the Entomological Society of Washington. 1995;**97**:779-790

Chapter 9

Diptera Development: A Forensic Science Perspective

Adrienne Brundage

Abstract

Insects, particularly Diptera, can reveal a great deal of information to investigators. By using known developmental data along with the common pattern of development displayed by flies, the time of colonization (and by logical extension, the postmortem interval) may be determined. This method requires investigators to know the exact development data of the insects at a scene and be able to do some simple calculations based on the concept of degree days (DD) or degree hours (DH). This chapter will give an overview of the methods currently used by forensic entomologists to translate the developmental cycle of flies into usable data for a crime scene.

Keywords: forensic entomology, Diptera, development, time of colonization, PMI

1. Introduction

Forensic entomology is the intersection between the study of arthropods and the justice system [1–4]. The study of arthropods, particularly insects and their close relatives, is entomology. The word comes from the Greek words "entomo," meaning insect, and "logus," meaning research. The area includes a variety of biological disciplines and areas of study, including forensics [5]. Such disciplines' common denominator is that they all include insects as the study subject.

A scene's analysis of entomological evidence has the potential to provide invaluable information about the scene and circumstances surrounding the situation. Interpretation of entomological evidence as the ability to inform an investigator about many different aspects of the case, including: colonization period and, by extension, death time; colonization season; colonization location; potential movement or storage of the remains after death; evidence of neglect; sites of trauma on the remains; and the presence of chemicals in the remains [3, 4, 6, 7]. Such data can then notify the investigation in several stages.

Insects reveal a lot to a knowledgeable researcher about a scene. First, the information most sought after is the time of colonization (TOC) estimate, which is often associated with the postmortem interval or time of death [1–4]. Most essential insects in forensics belong to the ecological class of decomposers—the ones that use the nutrients bound in dead matter [8]. Decomposers locate and exploit dead matter efficiently, and those who respond to ephemeral resources such as animal carcasses excel in the location of resources [8, 9]. Insects may arrive at a newly dead animal within minutes after death [7, 10] and either feed upon the carcass directly or colonize the carcass through egg oviposition [4, 8]. According to the resource position output, it can be concluded that an insect with unfettered access to a carcass arrived and colonized it within minutes of animal death. The calculation of the age of the insects colonizing the animal body (TOC estimate) can therefore be used as a measure of how long this carcass was available for insect colonization and, by logical extension, how long it was dead (postmortem interval or PMI) [4, 11].

2. The dipteran life cycle

Insect age estimation relies on their basic physiology. Insect are poikilothermic, meaning their growth and development is affected by ambient temperature [12]. Lower ambient temperatures lead to slower insect growth, while higher ambient temperatures lead to faster insect growth. Each insect species has an upper and lower threshold, above or below which the insect no longer grows [8, 13]. We can describe and predict this growth based on ambient temperature by using a mathematical formula [4, 14, 15].

Investigators are able to use the ability of insects to quickly locate and colonize carrion and the ability to mathematically estimate insect age given ambient temperature to estimate a postmortem interval, given some assumptions. If we assume that insects on an animal arrived right after death and that their growth was dictated by ambient temperature, then we can calculate how long it would take an observed insect to reach an observed developmental stage at a given ambient temperature. This calculated time is the postmortem interval [3, 4, 7]. If the exact arrival time of insects at carrion is unknown, however (i.e., before or after death, animal was blocked from insect activity, etc.) this calculated time is the time of colonization interval [4, 11].

There are two genders involved in reproduction in most species. It is the female reproductive system that produces and stores eggs, provides food for eggs, collects and stores sperm, fertilizes the eggs, and positions the eggs in the environment. The male reproductive system is responsible for sperm production and for egg fertilization presentation to the female [16].

The Diptera, or the true flies, are the main forensically significant insects. Because Diptera is typically the first insect to colonize animal remains, the most widely used in forensic cases is the dipteran life cycle. Therefore, the dipteran life cycle is considered the most important to understand [3, 7, 17]. All flies start as eggs normally laid in the environment, although some eggs remain in the female's body and are placed as larvae in the environment [18]. The adult flies enter a dead animal and lay their eggs or larvae on that animal's natural body openings and wounds. The resulting larvae feed on the body until it is ready for pupation. Flies have what is known as a complete life cycle or metamorphosis: beginning as embryos, hatching into larvae, pupating, and eclosing from the pupae into an adult. The larvae are in three stages or instars, before they finish feeding on the body and move into the dispersal or wandering stage. This dispersal stage takes the larvae 15–20 ft away from the carrion and to a protected place in which to pupate [8]. The dispersing larvae will crawl under things in the environment or dig a few centimeters down into the soil, where the outer cuticle of the larva hardens into a protective shell or pupal case. Inside this pupal casing, enzyme transforms the larvae into an adult [6].

3. Calculating time of colonization

Dipteran larvae are poikilothermic, so they do not generate their own body heat [5, 16, 19]. Rather, the ambient temperature affects them, which determines how rapidly or slowly they develop [12, 16, 19]. The warmer the temperature in the area, the faster they go through their instars and pupal stage. The higher the ambient temperature, the faster their instars and pupal stage are going through. This estate allows forensic entomologists to use information on insect growth to assess the time of colonization estimate of animal remains. This also makes temperature the most important density-independent factor for insects [4, 19].

Temperature has a profound effect on the metabolic of dipteran and the speed of development [16]. Warmer temperatures generally result in faster development, within certain temperature ranges. But, this is not accurate at extreme temperatures. Insects have both upper and lower thermal limits, below and above which the insect either dies or no longer develops. These are known as threshold temperatures [16, 19, 20]. In different species, these thermal levels are naturally different. Organisms that have evolved in tropical and warmer areas will have higher limits than those that have developed in temperate or colder areas [21]. Therefore, when it comes to temperature scales, there is a huge variation between species. The lower limits are better known than the upper limits and are significant when measuring a colonization estimate period mathematically [17, 20]. Developmental thresholds for forensically important flies are usually between 6 and 10°C and are experimentally determined. If a specific fly species developmental threshold is not available, a general rule of thumb is to use 6°C for winter species or flies in cold areas and 10°C for warm weather species or flies in warmer areas [7, 22–24].

Each stage of insect growth requires a certain amount of heat above the minimum temperature and below the maximum temperature, from egg to an adult [12, 25]. This growth rate can be represented with a linear model that represents the amount of heat necessary for a given fly to go through each of its developmental stages. These linear models are called degree day or degree hour models, because development is recorded as the temperature above the minimum developmental threshold multiplied by time (days or hours) [13, 14, 26, 27].

This model allows us to calculate the time a fly takes to develop using ambient temperature, and this formula is

$$(\text{average ambient temperature} - \text{minimum threshold}) \times \text{unit of time}$$

This formula takes the average ambient temperature over a given time period, subtracts the minimum threshold for the insect species, and multiplies the result by unit of time. The result is called a "degree day" (if the unit of time used was a day) or a "degree hour" (if the unit of time used was an hour). These are the two most common units of time seen in this type of model, simply because weather stations often record ambient temperature at daily or hourly intervals.

The formula may be used to calculate how many degree days or degree hours are necessary for an insect to get through various stages in its life cycle, as well as how many degree days have been accumulated over time. The DD/DH necessary for an insect to get through its lifecycle is determined by experiment. The data are recorded as hours needed for a species to develop at a given temperature, and these data may be converted into DD or DH simply by using the formula.

Table 1 shows several forensically important insects and the time it takes each insect to develop at various temperatures. Take, for example, the hours necessary for the black blow fly, *Phormia regina* (Meigen), to develop at 27°C. According to research, it takes *P. regina* eggs 16 h to go from freshly laid to hatching, 18 h to go through the first larval instar, 11 h to go through the second larval instar, and 36 h to go through the third larval instar. To turn this information into degree hours, simply plug the data into the equation

Egg DH:

$$(27°C - 10°C \times 16\ \text{h}$$
$$(17°C) \times 16\ \text{h}$$
$$272\ \text{DH}$$

In this case, it takes *P. regina* 272 DH to go through its entire egg stage. This same calculation may be used to determine DD:

Egg DD:

$$(27°C - 10°C \times 0.67\ \text{days}$$
$$(17°C) \times 0.67\ \text{days}$$
$$11.39\ \text{DD}$$

The number of DD or DH necessary for each stage of *P. regina* development may be calculated in this same way:

First instar DH/DD:

$$(27°C - 10°C \times 18\ \text{h}$$
$$(17°C) \times 18\ \text{h}$$
$$306\ \text{DH} \quad \text{or} \quad 12.75\ \text{DD}$$

Second instar DH/DD:

$$(27°C - 10°C \times 11\ \text{h}$$
$$(17°C) \times 11\ \text{h}$$
$$180\ \text{DH} \quad \text{or} \quad 7.5\ \text{DD}$$

Third instar DH/DD:

$$(27°C - 10°C \times 36\ \text{h}$$
$$(17°C) \times 36\ \text{h}$$
$$612\ \text{DH} \quad \text{or} \quad 25.5\ \text{DD}$$

The results may be summed up to determine the accumulated degree days (ADD) or accumulated degree hours (ADH) necessary for *P. regina* to develop from freshly oviposited egg through the end of the third instar:

$$11.39\ \text{DD} + 12.75\ \text{DD} + 7.5\ \text{DD} + 25.5\ \text{DD} = 57.14\ \text{DD}$$

This may be calculated for any range of developmental stages in an insect's life cycle.

Once the ADD or ADH necessary for insect development is known, it is possible to calculate the time necessary for the DD to accumulate at recorded ambient temperatures. The species threshold temperature is used in the same DD formula, but the average temperature is a weather station recording over a given period of time.

If, for example, a weather station recorded an average temperature of 15^{0}C over 1 day, and then the DD accumulated over that 1 day may be calculated as

$$(15°C - 10°C) \times 1\ \text{day}$$
$$(5°C) \times 1\ \text{day}$$
$$5\ \text{DD}$$

This calculation shows that 5 degree days accumulated over that 1 day. Each passing day may be calculated in the same way, and the results are added up to produce the accumulated degree days over any given period of time:

Average temperature (°C) on 4 separate days: 15°C, 20°C, 21°C, and 18°C
DD for each day: 5 DD, 10 DD, 11 DD, 8 DD
ADD over the 4 days: 5 DD + 10 DD + 11 DD + 8 DD = 34 DD

Once all this data is determined, the DD required for the growth of an insect species is added to the total DD accumulated over a period of time to decide how long it would take for an insect to develop into an observed stage of life given at the reported ambient temperature. It is best to display this data in table form:

Date	Average temp (°C)	DD	ADD
May 15	21	11	11
May 14	25	15	26
May 13	22	12	38
May 12	18	8	46
May 11	22	12	58
May 10	20	10	68
May 9	15	5	73

Table 1.
Example of average ambient temperatures and calculated DD/ADD from a weather station.

Note that the dates are described in descending order; this method allows an entomologist to begin with the date on which evidence of insects was collected and to work back in time to determine the minimum period of insect colonization. In **Table 1**, 11 DD were accumulated on May 15, 15 DD on May 14, 12 DD on May 13, etc. In total, it takes 57.14 DD to develop from egg through the end of the third instar.

If maggots at the very end of the third instar were found on an animal at the end of the day on May 15, the ADD necessary for those maggots to develop through the third instar may be used to determine when the eggs were laid. In this case, *P. regina* needs 57.14 DD to develop through the third instar. There were not enough DD accumulated on May 15 to reach that number, so the eggs could not have been laid on May 15 and have had time to reach the observed life stage. Since only 25 DD accumulated between May 14 and May 15, there was still not enough time for the fly to reach its third instar. Enough DD is accumulated between May 11 and May 15, however, for the insect to reach its observed life stage. In this instance, we can say the eggs had to have been laid sometime on May 11 or earlier to reach the third instar by May 15. This gives a minimum time of colonization estimation, based on the life history of the flies colonizing an animal.

The feasibility of this approach of the colonization estimation time degree day model depends on the availability of growth rate information for different necrophagous species. Such data are collected experimentally and appear to be reported as a life table or the rate of development of different insect species.

Arthropods associated with carrion are associated with ecological groups: necrophagous species, which feed upon the animal remains; omnivorous species, which feed upon both the animal remains and other organisms colonizing those remains; predators and parasites of the necrophagous or omnivorous species;

adventive species, which use the remains as an extension of the environment; and incidental species which are associated with animal remains due to happenstance [3, 28].

Necrophagous insects arrive on and in a corpse in a somewhat predictable sequence and are arguably the most useful for forensic entomology. This is the ecological succession of insects and is influenced by the environment, season, and the decomposition state of the carrion. The insects arrive in blending waves, called seres, of organisms, each comprised of different species attracted to a particular state of decay [3, 29].

Early research in entomology indicated that the number of seres varies according to the placement of the carrion. Mègnin [30] showed that animals exposed to the environment yielded eight distinct insect seres, while those that were buried attracted only three distinct insect seres. This difference speaks to the availability of the tissue for colonizing insects, along with the ability of those insects to reach tissue that is blocked from easy access. Different insects are attracted to different stages of generalized decomposition. While stages of decomposition are not discrete and can sometimes be difficult to characterize with precision, historically scientists have broken up decomposition of animal remains into five major stages: fresh, putrefaction, active decay, butyric fermentation, and dry decay [31]. Each of these stages attracts one or more seres of insects and other arthropods in a predictable sequence.

From the moment of death, the insect fauna of an animal body begins to change. Any ectoparasites associated with the body leave relatively quickly as the body cools and the blood ceases to circulate [18]. Myiasis-causing flies may or may not die—it is dependent upon if they are obligate parasites of living tissue or facultative parasites that can feed on both living and dead animals. Botflies, for example, are dependent upon a living host, and if the host dies, the botfly dies. On the other hand, *Cochliomyia* sp. can feed on living or dead hosts and may just continue to feed after the animal dies [4, 18].

Necrophagous insects or those attracted to dead tissue are attracted to the body within minutes of death and are associated with the fresh stage of decay. In general, the first adults are observed on a newly dead animal within an hour after death, as long as there is an adequate access to the body [4]. Some investigators report female flies arriving within 15 s of death [7]. Eggs and early instar maggots tend to appear with the onset of autolysis. In general, the animal will be characterized in this early sere by the presence of adult flies and fly eggs. Eggs will be found near natural bodily opening, on wounds, and sometimes in protected folds of skin or coverings. Those eggs will hatch to first instars quickly in warm environments, but there will not be a large maggot mass on the body; the tissue will still look fresh. The most common groups of insects found during this fresh stage include species from the fly families of Calliphoridae, Muscidae, and Sarcophagidae [32, 33].

As the carrion begins to putrefy, the young maggots will move throughout the body, spreading bacteria, secreting digestive enzymes, and feeding on tissue. They move as a maggot mass, benefitting from communal heat and shared digestive secretions. Larvae first feed between muscles then on the muscle fibers themselves as the maggots grow and the digestive juices get to work. The rate of decay increases, and the odors emitted from the body attract more blow flies, flesh flies, beetles, and mites. They are joined by parasitic wasps that lay their eggs inside maggots and later inside pupae. The most common insects associated with this putrefaction stage, however, are species in the fly families of Calliphoridae and Sarcophagidae [3].

Once the carrion enters in active decay, there will be several generations of maggots present on the body. Some of the maggots will be large well and enter into the third instar. The oldest maggots will begin to disperse from the carrion to find pupation sites. These maggots will crawl under objects in the environment or burrow into the soil to become pupae. The early seres or pioneer flies cease to be attracted to the corpse, making way for those insects that prefer later stages of decay. Predatory maggots are much more abundant at this stage and may be feeding upon other species in the maggot mass [34]. Predatory beetles lay their eggs in the corpse and their larvae then hatch to feed on the dead animal and other insects colonizing that animal. Parasitic wasps are even more common, attacking the huge maggot masses and developing pupae. Active decay attracts a new sere characterized by beetles in Dermestidae family and grease moths. As active decay moves into butyric fermentation, two additional seres are attracted: first, a sere consisting of cheese skippers, *Fannia* sp.; Silphid beetles; and beetles in the family Cleridae arrives at the body. The second sere consists of dump flies, flies in the family Phoridae, beetles in the family Silphidae, and clown beetles [3].

As the carrion is stripped of most of its soft tissue, the remaining tissue begins to dry out. The reduction in soft food makes the body less palatable to the mouth hooks of maggots and more suitable for the chewing mouthparts of beetles. Beetle adults and larvae feed on skin and ligaments, while certain late stage flies, such as the cheese fly, arrive to feed on whatever moist flesh remains. Predators and parasites are still prevalent at this stage, feeding on the straggling maggots or other soft-bodied insects in the vicinity [3, 35].

Once the carrion has completely dried out, the body attracts insects that can feed on hair and dried skin. This dry stage of decomposition is highly attractive to a new sere consisting of mites, followed by a sere of beetles in Dermestidae and Tingidae, and finally a sere of primarily Ptinidae and Tenebrionidae. This stage tends to last for a long while, since the business of feeding on dried out tissue takes a great deal of time and digestive enzymes. The beetles will remain on the bones as long as the carrion is undisturbed and has dried tissue [36].

After the dried tissue has been cleaned by the beetles, moths, and mites, only the bones remain along with empty pupal casings. There is no longer any major insect activity, and any further decomposition is accomplished by bacteria and physical factors [31].

By way of comparison, a buried animal has fewer seres with lower species diversity [37]. Fresh buried animals tend to attract flies in Calliphoridae, Muscidae, Sarcophagidae, and Phoridae. Buried animals in active decay may attract root-eating beetles, and those in dry decay are populated by rove beetles.

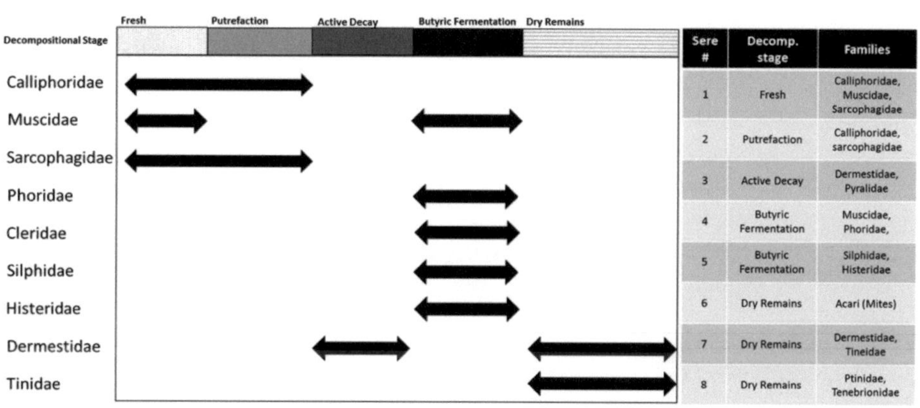

Figure 1.
General arrival of forensically important insect families by phase of decomposition.

Since Meginin's original experiments, there have been many, many attempts to characterize the number of seres that show up on animals of different sizes. The consensus is that these successional waves range from 8 to 10, and any attempts to discreetly define the arthropod community associated with the seres is confounded by the continual nature of decomposition. There is a broad general agreement of orders and families that show up on a decomposing body [3, 4, 7]. There is also a general agreement of sequence and the idea that the first sere is primarily composed of Dipteran species. Succession is affected by the location of the carrion, any covering, animal type, local insect species, time of year, and a myriad of other factors. At the family level, the succession looks like as shown in **Figure 1**. Knowing the general succession of insects on a corpse in a particular area enables the entomologist to extend the time of colonization estimation even if the primary colonizers of the first sere have come and gone.

4. Conclusions

Diptera are currently the most important insect when it comes to forensic entomology. Detailed knowledge of common flies colonizing a body and the time taken by those flies to develop is the bedrock of the science. While much is known about this work, there is always much more to do.

Author details

Adrienne Brundage
Texas A&M - Agriculture and Life Sciences, Texas A&M University, College Station, USA

*Address all correspondence to: brundage@forensicentomologist.com

References

[1] Amendt J et al. Best practice in forensic entomology—Standards and guidelines. International Journal of Legal Medicine. 2007;**121**:90-104

[2] Benecke M. A brief history of forensic entomology. Forensic Science International. 2001;**120**:2-14

[3] Smith KG. A Manual of Forensic Entomology. Oxford: British Museum of Natural History, Cornell University Press, University Printing House; 1986

[4] Byrd JH, Castner JL. In: Byrd JH, Castner JL, editors. Forensic Entomology: The Utility of Arthropods in Legal Investigations. 2nd ed. Boca Raton: Tayler & Francis; 2010. xxiii, 681 p

[5] Gullan PJ, Cranston PS. The Insects: An Outline of Entomology. Chichester, West Sussex, UK: John Wiley & Sons Ltd.; 2009

[6] Barnes KM, Gennard DE. The effect of bacterially-dense environments on the development and immune defences of the blowfly *Lucilia sericata*. Physiological Entomology. 2011;**36**(1): 96-100

[7] Catts EP, Haskell NH. In: Catts EP, Haskell NH, editors. Entomology and Death: A Procedural Guide. 2nd ed. Clemson, SC: Joyce's Print Shop, Inc.; 2008. xii, 182 p

[8] Price PW. Insect Ecology. Chichester, West Sussex, UK: John Wiley & Sons Ltd.; 1997

[9] Cain ML, Bowman WD, Hacker SD. In: Cain ML, editor. Ecology. Sunderland: Sinauer Associates, Inc.; 2008. p. 621

[10] Hall MJM. Trapping the flies that cause myiasis: Their responses to host-stimuli. Annals of Tropical Medicine and Parasitology. 1995;**89**(4):333-357

[11] Tomberlin JK et al. A roadmap for bridging basic and applied research in forensic entomology. Annual Review of Entomology. 2011;**56**(1):401-421

[12] Davidson J. On the relationship between temperature and rate of development of insects at constant temperatures. The Journal of Animal Ecology. 1944;**13**(1):26-38

[13] Yang S, Logan J, Coffey DL. Mathematical formulae for calculating the base temperature for growing degree days. Agricultural and Forest Meteorology. 1995;**74**(1–2):61-74

[14] Michaud J-P, Moreau G. A statistical approach based on accumulated degree-days to predict decomposition-related processes in forensic studies. Journal of Forensic Sciences. 2011;**56**(1):229-232

[15] Higley LG, Pedigo LP, Ostlie KR. DEGDAY: A program for calculating degree-days, and assumptions behind the degree-day approach. Environmental Entomology. 1986;**15**(5): 999-1016

[16] Nation JL. Insect Physiology and Biochemistry. CRC Press; 2011

[17] Frost CL et al. Indoor arthropods of forensic importance: Insects associated with indoor decomposition and mites as indoor markers. In: Current Concepts in Forensic Entomology. Netherlands: Springer; 2010. pp. 93-108

[18] Mullen GR, Durden LA. Medical and Veterinary Entomology. Academic Press; 2002

[19] Beck SD. Insect thermoperiodism. Annual Review of Entomology. 1983; **28**(1):91-108

[20] Block W. Cold hardiness in invertebrate poikilotherms. Comparative Biochemistry and

Physiology Part A: Physiology. 1982; **73**(4):581-593

[21] Addo-Bediako A, Chown SL, Gaston KJ. Thermal tolerance, climatic variability and latitude. Proceedings of the Royal Society of London. Series B: Biological Sciences. 2000;**267**(1445): 739-745

[22] Ames C, Turner B. Low temperature episodes in development of blowflies: implications for postmortem interval estimation. Medical & Veterinary Entomology. 2003;**17**:178-186

[23] Davies L, Ratcliffe GG. Development rates of some pre-adult stages in blowflies with reference to low temperatures. Medical and Veterinary Entomology. 1994;**8**:245-254

[24] Anderson GS. Minimum and maximum development rates of some forensically important Calliphoridae (Diptera). Journal of Forensic Sciences. 2000;**45**(4):824-832

[25] Hagstrum DW, Milliken GA. Quantitative analysis of temperature, moisture, and diet factors affecting insect development. Annals of the Entomological Society of America. 1988; **81**(4):539-546

[26] Pruess KP. Day-degree methods for pest management. Environmental Entomology. 1983;**12**(3):613-619

[27] Megyesi MS, Nawrocki SP, Haskell NH. Using accumulated degree-days to estimate the postmortem interval from decomposed human remains. Journal of Forensic Sciences. 2005;**50**(3):618-626

[28] Roberts J, Márquez-Grant N. Forensic Ecology Handbook: From Crime Scene to Court. Vol. 9. John Wiley & Sons; 2012

[29] Goff ML. Estimation of post-mortem interval using arthropods' development and successional patterns. Forensic Science Reivew. 1993;**5**:81-94

[30] Mégnin P. La Faune des cadavres: Application de l'entomologie à la médicine légale. Masson & Gauthier-Villars; 1894

[31] Haglund WD, Sorg MH. Forensic Taphonomy: The Postmortem Fate of Human Remains. CRC Press; 1996

[32] Smith PH. Causes and correlates of loss and recovery of sexual receptivity in *Lucilia cuprina* females after their 1st mating. Journal of Insect Behavior. 1989;**2**(3):325-337

[33] Schoenly K, Reid W. Dynamics of heterotrophic succession in carrion arthropod assemblages: Discrete seres or a continuum of change? Oecologia. 1987; **73**(2):192-202

[34] Nelder MP, McCreadie JW, Major CS. Blow flies visiting decaying alligators: Is succession synchronous or asynchronous? Psyche. 2009:1-7

[35] Schoenly K. A statistical analysis of successional patterns in carrion-arthropod assemblages: Implications for forensic entomology and determination of the postmortem interval. Journal of Forensic Sciences. 1992;**37**(6):1489-1513

[36] Szpila K. Key for the identification of third instars of European blowflies (Diptera: Calliphoridae) of forensic importance. In: Current Concepts in Forensic Entomology. Netherlands: Springer; 2010. pp. 43-56

[37] Payne JA. Arthropod succession and decomposition of burried pigs. Nature. 1968;**219**(5159):1180-1968